AF413051

# CIRCUIT SYNTHESIS WITH VHDL

# THE KLUWER INTERNATIONAL SERIES IN ENGINEERING AND COMPUTER SCIENCE

## VLSI, COMPUTER ARCHITECTURE AND DIGITAL SIGNAL PROCESSING

*Consulting Editor*
**Jonathan Allen**

## *Other books in the series:*

**HOT-CARRIER RELIABILITY OF MOS VLSI CIRCUITS**, Y. Leblebici, S. Kang
ISBN: 0-7923-9352-X
**MOTION ANALYSIS AND IMAGE SEQUENCE PROCESSING**, M. I. Sezan, R. Lagendijk
ISBN: 0-7923-9329-5
**HIGH-LEVEL SYNTHESIS FOR REAL-TIME DIGITAL SIGNAL PROCESSING: The Cathedral-II Silicon Compiler**, J. Vanhoof, K. van Rompaey, I. Bolsens, G. Gossens, H. DeMan
ISBN: 0-7923-9313-9
**SIGMA DELTA MODULATORS: Nonlinear Decoding Algorithms and Stability Analysis**, S. Hein, A. Zakhor
ISBN: 0-7923-9309-0
**LOGIC SYNTHESIS AND OPTIMIZATION**, T. Sasao
ISBN: 0-7923-9308-2
**ACOUSTICAL AND ENVIRONMENTAL ROBUSTNESS IN AUTOMATIC SPEECH RECOGNITION**, A. Acero
ISBN: 0-7923-9284-1
**DESIGN AUTOMATION FOR TIMING-DRIVEN LAYOUT SYNTHESIS**, S. S. Sapatnekar, S. Kang
ISBN: 0-7923-9281-7
**DIGITAL BiCMOS INTEGRATED CIRCUIT DESIGN**, S. H. K. Embadi, A. Bellaouar, M. I. Elmasry
ISBN: 0-7923-9276-0
**WAVELET THEORY AND ITS APPLICATIONS**, R. K. Young
ISBN: 0-7923-9271-X
**VHDL FOR SIMULATION, SYNTHESIS AND FORMAL PROOFS OF HARDWARE**, J. Mermet
ISBN: 0-7923-9253-1
**ELECTRONIC CAD FRAMEWORKS**, T. J. Barnes, D. Harrison, A. R. Newton, R. L. Spickelmier
ISBN: 0-7923-9252-3
**ANATOMY OF A SILICON COMPILER**, R. W. Brodersen
ISBN: 0-7923-9249-3
**FIELD-PROGRAMMABLE GATE ARRAYS**, S. D. Brown, R. J. Francis, J. Rose, S. G. Vranesic
ISBN: 0-7923-9248-5
**THE SECD MICROPROCESSOR, A VERIFICATION CASE STUDY**, B. T. Graham
ISBN: 0-7923-9245-0
**HIGH LEVEL SYNTHESIS OF ASICs UNDER TIMING AND SYNCHRONIZATION CONSTRAINTS**, D. C. Ku, G. De Micheli
ISBN: 0-7923-9244-2
**FAULT COVERING PROBLEMS IN RECONFIGURABLE VLSI SYSTEMS**, R. Libeskind-Hadas, N. Hassan, J. Cong, P. McKinley, C. L. Liu
ISBN: 0-7923-9231-0
**VHDL DESIGNER'S REFERENCE**, J-M. Bergé, A. Fonkoua, S. Maginot, J. Rouillard
ISBN: 0-7923-1756-4
**A FORMAL APPROACH TO HARDWARE DESIGN**, J. Staunstrup
ISBN: 0-7923-9427-5

# Circuit Synthesis with VHDL

by

ROLAND AIRIAU,
JEAN-MICHEL BERGE
AND VINCENT OLIVE
*CNET, France Telecom, France*

**KLUWER ACADEMIC PUBLISHERS**
BOSTON / DORDRECHT / LONDON

Library of Congress Cataloging-in-Publication Data

```
Airiau, Roland.
   Circuit synthesis with VHDL / by Roland Airiau, Jean.Michel Bergé,
Vincent Olive.
      p.   cm. -- (The Kluwer international series in engineering and
computer science ; v. 261)
   Includes index.
   ISBN 0-7923-9429-1 (hb : acid-free paper)
   1. VHDL (Computer hardware description language) 2. Logic design-
-Data processing. 3. Computer-aided design.   I. Bergé, Jean
-Michel.  II. Olive, Vincent.  III. Title.  IV. Series: Kluwer
international series in engineering and computer science ; SECS 261.
TK7885.7.A37   1994
621.39'5--dc20                                        93-42978
```

ISBN 0-7923-9429-1


Published by Kluwer Academic Publishers,
P.O. Box 17, 3300 AA Dordrecht, The Netherlands.

Kluwer Academic Publishers incorporates
the publishing programmes of
D. Reidel, Martinus Nijhoff, Dr W. Junk and MTP Press.

Sold and distributed in the U.S.A. and Canada
by Kluwer Academic Publishers,
101 Philip Drive, Norwell, MA 02061, U.S.A.

In all other countries, sold and distributed
by Kluwer Academic Publishers Group,
P.O. Box 322, 3300 AH Dordrecht, The Netherlands.

*Printed on acid-free paper*



Printed in the Netherlands

A designer, his boss, and one of the VHDL language designers are in the same train traveling through Scotland. Suddenly, through the window, they catch sight of a black ship.

"Oh," says the boss, "In Scotland ships are black."

"No," says the designer, "All we can say is that in Scotland, at least one ship is black."

"I am sorry ," interrupts the VHDL language designer, "The only thing we can say is that in Scotland, there is at least one ship one side of which is black."

# TABLE OF CONTENTS

# LIST OF FIGURES

# *Acknowledgments*

*"Circuit Synthesis with VHDL"* has been made possible thanks to the support we have received from Jean-Pierre Noblanc, Jean-Louis Lardy and Jacques Lecourvoisier at the Centre National d'Etudes des Télécommunications[1] (France Telecom).

We greatly appreciated the preliminary reviews of  Serge Maginot, Anne Robert, Jacques Rouillard and Denis Rouquier and, the very useful comments given by Maureen Timmins and Jeanine Jenks.

We would especially like acknowledge
* the CNET which has a long tradition in standardization,
* the ECIP project which groups together almost all standardization efforts in Europe,
* all our friends, colleagues and relations who are participating to the European VHDL synthesis working group and IEEE standardization working groups such as the SSIG (Synthesis Special Interest Group). We know that the results of these groups will soon be accepted and appreciated by the entire VHDL community.

---

[1] CNET, 28 Chemin du Vieux Chêne, BP 98, 38243 Meylan Cedex — France

# FOREWORD

This book is an introduction to the use of VHDL synthesis tools in design. Logic (RTL) synthesis and some aspects of architectural synthesis are developed in it.

### Modeling styles

The proposed modeling style does not target a specific market tool. Since there is no standard modeling style, this book is a tradeoff between what the designer could legitimately expect from the tools and what a synthesis tool builder can reasonably implement.

Therefore, some of the constructs presented in this book may not be synthesizable on all synthesis tools. The following pages focus on constructs universally recognized as synthesizable by synthesis tools. Nevertheless, in some cases, the designer is warned that a given construct, even if conceptually synthesizable, is not currently recognized by all synthesis tools.

Of course, this classification is a snapshot of a moving situation. The designers, as well as the authors, are expecting a convergence to a common set of recognized constructs to occur soon.

### VHDL'92

A new version of VHDL, commonly referenced as *VHDL'92*, was accepted in September 1993 and is, at least theoretically, the only legal version of VHDL. This version of the language is upward compatible with the old one, sometimes referred to as *VHDL'87*. Indeed, the new version has no major consequences on the modeling style for synthesis. Nevertheless, when certain new constructs appear to be used in the synthesis domain, they may be illustrated, and signaled as such, in this book. Since all synthesis tools are far from accepting all new constructs, a *VHDL'87* version of the same example is always provided. For more details on *VHDL'92*, please refer to [BER92].[2]

---

[2] J. M. Bergé, A. Fonkoua, S. Maginot, J. Rouillard, "VHDL'92: The New Features of the VHDL Hardware Description Language," Kluwer Academic Publishers, June 1993.

### *Architectural Synthesis*

This book mainly focuses on the logic synthesis process, which is the only one really proposed by tool vendors. The difference between logic and architectural synthesis mainly depends on the intelligence of the tools and not on the descriptive capabilities of the language. This publication may offer the designer the modeling capabilities of entering the architectural synthesis process.

### *Combinational/Sequential and Synchronous/Asynchronous*

Throughout this book, the words *combinational*, *sequential*, *synchronous*, and *asynchronous* are used for VHDL descriptions. Without entering philosophical considerations, the following simplified definitions are sufficient for a good understanding of the book.

*Combinational/sequential* are hardware notions:
- A VHDL description is said to be *combinational* if its inferred hardware does not involve any memorization elements.
- A VHDL description, that is not combinational is said to be *sequential*.

*Synchronous/asynchronous* are abstractions of system-level:
- A VHDL description is said to be *synchronous* if a specific control signal called a clock is present. If all synchronization points rely on this clock, the design is purely synchronous. Since synchronous implies memorization, synchronous designs, either purely or not, are sequential.
- The word *asynchronous* is used in this book to characterize parts of the VHDL description that do not rely on a clock. Combinational circuits are therefore asynchronous, but this does not necessary deny memorization (e.g., asynchronous FIFO).

# ERRATUM

Architectures SYNTH_LEVEL and SECOND_SYNTH_LEVEL (Pages 200 and 203) are incorrect. The authors apologize for this and would ask you to take this ERRATUM into account. The incorrect lines in the original version are preceded by the symbol # in this text.

```
architecture SYNTH_LEVEL of TRAFFIC is
#       constant PERIOD : TIME := 1 sec;            -- Clock period (Assuming type TIME accepted by the synthesis tool)
#       constant HALF_PERIOD :TIME:= PERIOD / 2;                -- Half clock period
#       constant LONG_TIME_SEC :NATURAL:= LONG_TIME / PERIOD;        -- Long_time in clock period units
#       constant SHORT_TIME_SEC : NATURAL := SHORT_TIME / PERIOD;
        signal CLK : BIT;                                       -- Local clock signal
        signal CURRENT_STATE, NEW_STATE : T_LIGHT; -- Intermediate signals coding  current and new state
        signal PRIORITY : BOOLEAN;                  -- Memorizes the edge of the external signal PRIORITY_MODE
        signal NO_PRIORITY : BOOLEAN;                   -- Priority has been taken in account
#       signal LG_TIME : NATURAL range 0 to  LONG_TIME_SEC;      -- Long timer
#       signal SH_TIME : NATURAL range 0 to SHORT_TIME_SEC;     -- Short timer
        signal LD_LG_TIME, LD_SH_TIME : BOOLEAN;            -- Timer load orders
        signal DEC_LG_TIME, DEC_SH_TIME : BOOLEAN;          -- Timer decrement orders
begin
        CLOCK_GENERATOR : CLK <= not CLK after HALF_PERIOD;     -- not synthesizable statement
        STATE_REGISTER : process (CLK, ECONOMY_MODE)           -- state register to be synthesized
        begin
                if ECONOMY_MODE then CURRENT_STATE <= FLASHING_RED;
                elsif CLK='1' and CLK'event then CURRENT_STATE <= NEW_STATE;
                end if;
        end process STATE_REGISTER;
        -- The CROSSROAD_LIGHTS signal reflects the internal values of the CURRENT_STATE signal
        CROSSROAD_LIGHTS <= CURRENT_STATE;
        PRIORITY_MEMO : process (CLK, ECONOMY_MODE) -- To translate the rising edge of PRIORITY signal to a level
        begin
                if ECONOMY_MODE then PRIORITY <= FALSE;
                elsif CLK='1' and CLK'event then
                        if PRIORITY_MODE then PRIORITY <= TRUE;
                        elsif NO_PRIORITY then PRIORITY <= FALSE;
                        end if;
                end if;
        end process;
        TIMERS : process (CLK) -- Synchronous decrementers as timers
        begin
                if LD_LG_TIME then LG_TIME <= LONG_TIME_SEC;
                elsif DEC_LG_TIME and (LG_TIME /= 0) then LG_TIME <= LG_TIME - 1;
                end if;
                if LD_SH_TIME then SH_TIME <= SHORT_TIME_SEC;
                elsif DEC_SH_TIME and (SH_TIME /= 0) then SH_TIME <= SH_TIME - 1;
                end if;
        end process ;
        -- Combinational part computing the new state and resetting the eventually memorized priority state
        COMPUTE_NEW_STATE : process (CURRENT_STATE, ECONOMY_MODE,  PRIORITY, CAR_DETECTOR)
        begin
                -- Initial values
                if ECONOMY_MODE then NO_PRIORITY <= FALSE; NEW_STATE <= FLASHING_RED;
                else
                        -- Default values of signals
                        NEW_STATE <= CURRENT_STATE;     NO_PRIORITY <= PRIORITY;    LD_LG_TIME <= FALSE;
                        LD_SH_TIME <= FALSE;            DEC_LG_TIME <= FALSE;       DEC_SH_TIME <= FALSE;
                        case CURRENT_STATE is
                        when FLASHING_RED =>
                                NEW_STATE <= MAIN_ROAD_GREEN; LD_LG_TIME <= TRUE;
                        when MAIN_ROAD_GREEN =>
                                if PRIORITY then LD_LG_TIME <= TRUE;
                                        elsif LG_TIME = 0 then
                                        if CAR_DETECTOR = CAR then
                                                NEW_STATE <= MAIN_ROAD_YELLOW; LD_SH_TIME <= TRUE;
                                        end if;
                                        else DEC_LG_TIME <= TRUE;
```

```vhdl
                    end if;
            when MAIN_ROAD_YELLOW =>
                if PRIORITY then NEW_STATE <= MAIN_ROAD_GREEN; LD_LG_TIME <= TRUE;
                elsif SH_TIME = 0 then NEW_STATE <= FARM_ROAD_GREEN; LD_LG_TIME <= TRUE;
                else DEC_SH_TIME <= TRUE;
                end if;
            when FARM_ROAD_GREEN =>
                if LG_TIME = 0 then NEW_STATE <= FARM_ROAD_YELLOW ; LD_SH_TIME <= TRUE;
                else DEC_LG_TIME <= TRUE;
                end if;
            when FARM_ROAD_YELLOW =>
                if SH_TIME = 0 then NEW_STATE <= MAIN_ROAD_GREEN; LD_LG_TIME <= TRUE;
                else DEC_SH_TIME <= TRUE;
                end if;
            end case ;
        end if;
    end process;
end SYNTH_LEVEL ;


architecture SECOND_SYNTH_LEVEL of TRAFFIC is
#       constant PERIOD : TIME := 1 sec;            -- Clock period (Assuming type TIME accepted by the synthesis tool)
#       constant HALF_PERIOD :TIME:= PERIOD / 2;                    -- Half clock period
#       constant LONG_TIME_SEC : NATURAL:= LONG_TIME / PERIOD;      -- Long_time in clock period units
#       constant SHORT_TIME_SEC : NATURAL:= SHORT_TIME / PERIOD;
        signal CLK : BIT;                                          -- Local clock signal
        signal PRIORITY : BOOLEAN;        -- Memorizes the edge of the external signal PRIORITY_MODE
begin
        CLOCK_GENERATOR : CLK <= not CLK after HALF_PERIOD;    -- Not synthesizable statement
        FSM : process (CLK, ECONOMY_MODE)    -- Process describing the Final State Machine and the operative part
            variable STATE : T_LIGHT;                              -- State register
#           variable LG_TIME : NATURAL range 0 to LONG_TIME_SEC;  -- Long timer
#           variable SH_TIME : NATURAL range 0 to SHORT_TIME_SEC; -- Short timer
        begin
            if ECONOMY_MODE then STATE := FLASHING_RED;                     -- Asynchronous part
            elsif CLK'event and CLK='1' then PRIORITY <= PRIORITY_MODE;    -- Synchronous part
                case STATE is
                when FLASHING_RED =>
                    STATE := MAIN_ROAD_GREEN;  LG_TIME := LONG_TIME_SEC;
                when MAIN_ROAD_GREEN =>
                    if PRIORITY then
                        LG_TIME := LONG_TIME_SEC;   PRIORITY <= FALSE;
                    elsif LG_TIME = 0 then
                        if CAR_DETECTOR = CAR then STATE := MAIN_ROAD_YELLOW;
                            SH_TIME := SHORT_TIME_SEC;
                        end if;
                    else LG_TIME := LG_TIME -1;
                    end if;
                when MAIN_ROAD_YELLOW =>
                    if PRIORITY then STATE := MAIN_ROAD_GREEN;
#                       LG_TIME := LONG_TIME_SEC;
                        PRIORITY <= FALSE;
                    elsif SH_TIME = 0 then STATE := FARM_ROAD_GREEN; LG_TIME :=LONG_TIME_SEC;
                    else SH_TIME := SH_TIME - 1;
                    end if;
                when FARM_ROAD_GREEN =>
                    if LG_TIME = 0 then STATE := FARM_ROAD_YELLOW ; SH_TIME := SHORT_TIME_SEC;
#                   else LG_TIME := LG_TIME - 1;
                    end if;
                when FARM_ROAD_YELLOW =>
                    if SH_TIME = 0 then    STATE := MAIN_ROAD_GREEN; LG_TIME := LONG_TIME_SEC;
                    else SH_TIME := SH_TIME - 1;
                    end if;
                end case ;
            end if;
            -- The signal CROSSROAD_LIGHTS reflects the internal values of the variable STATE
            CROSSROAD_LIGHTS <= STATE;
        end process;
end SECOND_SYNTH_LEVEL;
```

# 1. ABOUT SYNTHESIS

## 1.1. WHY VHDL?

A new era is opening up in electronics system design. By analogy with software programmers who no longer have to deal with internal coding details of computers, electronics designers are less and less concerned with circuit layout and standard components, especially when programmable ones (FPGA, EPLD) are targeted. New synthesis techniques are raising the level of abstraction of electronic device descriptions. By using the latter, the designer can deal with more complex applications with a complete mastery of the technologies. Furthermore, such techniques allow him or her to remain apart from the explicit description of the physical level. Abstraction means symbolic coding and therefore language, while synthesis techniques owe their wide use to hardware description languages.

Chronologically speaking, VHDL is not the first hardware description language. Created to be a standard, it is now widely accepted and adopted. VHDL owes its success both to its predecessors and to the maturity of its basic principles. It is the result of numerous studies on logic simulation principles and associated languages. Through this connection, it inherits from the electronic world.

VHDL language designers have taken a sufficiently objective stance in order to remain apart from the electronic domain. In this too narrow context of electronics, the designed language could not skip a generation, which VHDL does. Above all, a purely electronic language would have had little chance of becoming a reference standard: competitors are numerous.

The combination of computing and electronics is the key to the success of VHDL. In the software domain, languages have acquired a high degree of maturity. VHDL represents the association of the most up-to-date concepts with the specificities of hardware description languages.

The power and richness of the language do in fact overtake the original scope of VHDL. VHDL is a general purpose simulation language, and electronics may be seen as a given application domain. Other technologies (optics, pneumatics, etc.) are potentially concerned. New domains such as industrial process simulation, neuronics, or networks are also covered. The only real restriction to the use of VHDL for a given application is that the algorithm of simulation must be event-driven, which means that actions occur at discrete points in time and may imply other actions at discrete points in time. In the future, its extension to analog systems may deal with continuous systems: a standardization process is in progress.

## 1.2. VHDL FOR WHICH PURPOSE?

Simulation (i.e., execution of a description in order to check its time behavior) is not the only possible application of hardware description languages. Other applications such as synthesis already exist:

- formal proof, which is a technique allowing comparison of two descriptions of different levels. This proof is performed according to specific criteria and these techniques are promising. They potentially allow, at least in some cases, the certification that a low-level description is consistent with a functional description. No simulation will be performed for this, but the result will be a real certification and not only a validation as provided by a conventional simulation (i.e., only given a specific set of stimuli).
- documentation that allows hardware description language to be used to describe a system in an informational way: to explain the implementation choices in a nonambiguous and standard manner. This description is potentially a part of the design documentation and should offer enough abstraction to allow, if necessary, another system design in another technology.
- specification that consists in using a high-level language to state the requirements in an unambiguous way. Such a modeling should not only describe the desired system behavior but also its functional and timing constraints.

This book focuses on another important application domain of hardware description languages: synthesis. Its goal is to propose to designers the bases of a new methodology as well as to give them a thorough understanding of VHDL for this specific purpose. On the other hand, the reader will only find here a quick overview of the fundamental techniques of synthesis, just enough for a good mastery of the results. *Circuit Synthesis with VHDL* is entirely devoted to the modeling techniques for synthesis as well as to the setting up of the underlying design methodology.

## 1.3. IS VHDL A GOOD LANGUAGE FOR SYNTHESIS?

The language VHDL was not originally designed for synthesis. All VHDL constructs are not synthesizable: access types or files for instance have no direct hardware correspondence. However, it is perfectly possible to design a language in which all constructs are synthesizable. UDL/I, the Japanese Standard[3][4] proves this. Therefore, is VHDL unsuited to synthesis?

To negatively answer this question, it is first of all fundamental to really understand that the VHDL modeling for synthesis of a system is usually associated with its testbench. The testbench is the part of the description that models the environment of the system and allows its testing. This separation between system and testbench is important. Once the system is synthesized, it is possible to reuse the same testbench to simulate the result of the synthesis process and check its consistency. Figure 1.1 illustrates this methodology.

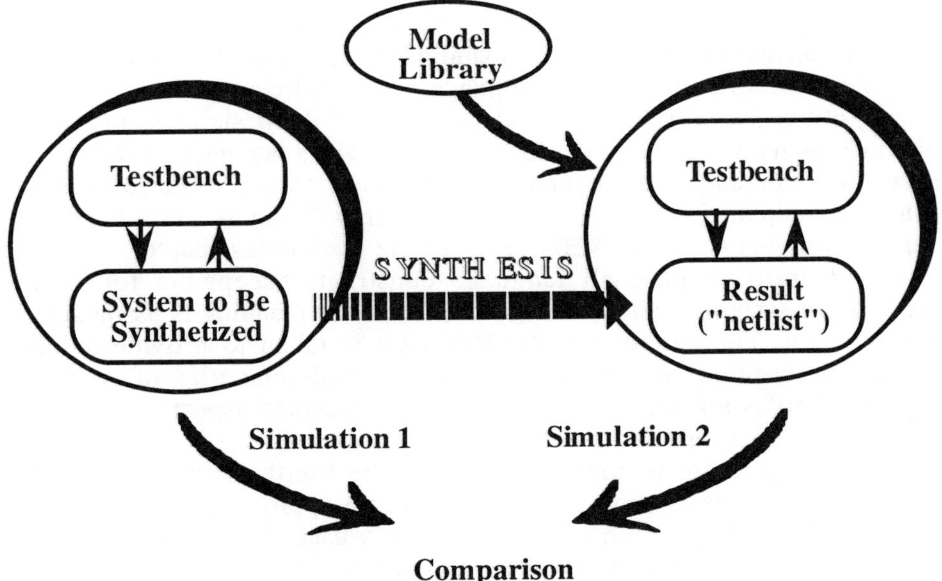


**Figure 1.1. Methodology for Synthesis Result Validation**

[3] "UDL/I Language Reference", Japan Electronic Industry Development Association

Often seen as a simple collection of stimuli, the testbench description can be complex, and multiple scenarios may be involved. A complexity of the same order as that of the system is not exceptional. All the power of VHDL can be used when describing the testbench: it will not be synthesized.

Furthermore, synthesis techniques are in their early stages and should evolve dramatically in the next few years. It is therefore essential that the language widely adopted for synthesis will in no way block progress to new levels of synthesis and increasing abstractions. The generality of VHDL allows it to deal with any synthesis level whatever its abstraction.


## 1.4. A BOOK, AN OUTLINE

How is this book made up? What path should one follow to achieve a good understanding and use of synthesis techniques? Figure 1.2 summarizes this.

In this introductory chapter, synthesis is explained in order to extract its clear definition, to show its inherent advantages, to determine the different classes of synthesis as well as their domains and limits of use, to understand the basic techniques, and finally to evaluate the consequences of such techniques on the design cycle.

In the second chapter, VHDL is studied in a more global way. It is not an academic in-depth course on the language, but a sufficiently detailed overview to bring out and achieve a good understanding of the basic concepts and their relationships with the synthesis domain. If VHDL is becoming *THE* hardware description language of so many tool vendors (often used for input and sometimes output of synthesis tools), it is mainly thanks to these concepts. Of course, a designer skilled in VHDL might want to skip this chapter.

In the third chapter, all language constructs acceptable for synthesis purposes are systematically detailed. The hardware meaning is associated with each of them. Throughout this chapter, care is taken to provide a full and accurate description of the synthesis syntax and semantics offered by the language. This is not restricted to present-day accepted aspects; "avant-garde" language constructs are also given and signaled.

As a complement to the previous one, the fourth chapter addresses the modeling-for-synthesis problem in a bottom-up approach. This chapter goes through the main target architectures commonly used. Modeling style advice is given and each problem is illustrated with examples.

Chapter five discusses the necessary underlying methodology when modeling for synthesis purposes. Which level of abstraction should one choose?

---

[4] J. M. Bergé, A. Fonkoua, S. Maginot, J. Rouillard, "VHDL Designer's Reference," Kluwer Academic Publishers, June 1992

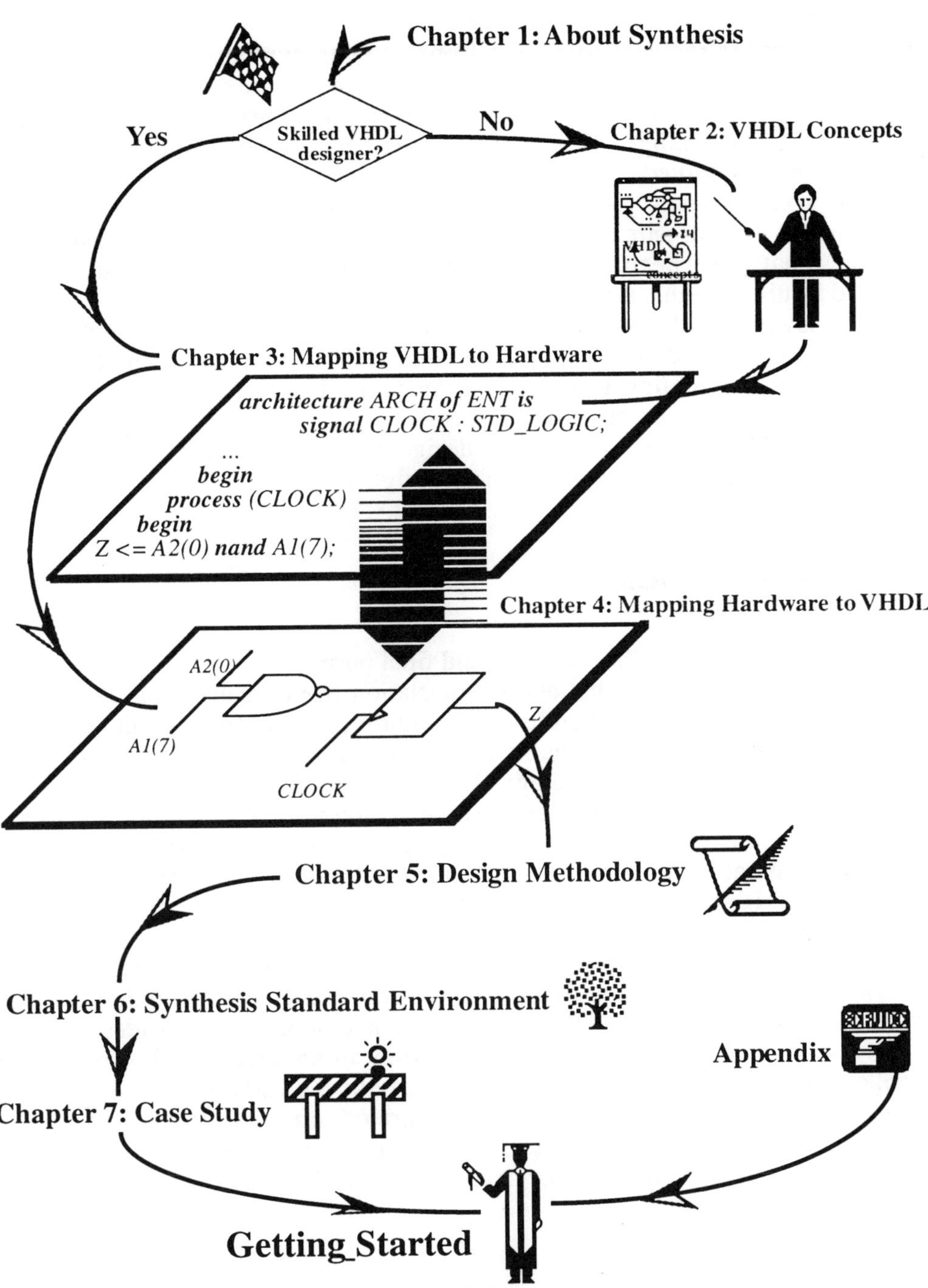


Figure 1.2. Synopsis of the Book

Which degree of detail is it necessary to achieve in the architectural choices? What is realistic to expect from existing synthesis tools? Since the functional aspects of synthesis have been solved in previous chapters, this chapter can answer these fundamental questions. Furthermore, interaction between user and synthesis tool, such as parameters for controling the synthesis process, are also studied.

The sixth chapter presents the synthesis standard environment. This environment is not part of the language but appears as separate standard packages that have been (or will soon be) balloted. Some conventions and standard synthesis semantics are also defined.

The seventh and last chapter puts into practice the different modeling techniques on a real case of significant complexity. The goal of this chapter is to match basic modeling techniques against the functional, topological, and timing requirements of a real-life problem.

Finally, the appendix allows the designer to quickly access much useful information on providing a synthesizable design style.

## 1.5. SYNTHESIS  DOMAIN

What is synthesis? What is its use and final purpose?

Synthesis is a process that allows translation from a behavioral description to a structural description where each element represents (or can automatically generate) a predefined electronic resource. The original functionality has to be maintained, and timing and topological requirements have to be met.

Etymologically speaking, *analysis* or *projection* would be closer to the concept than the word *synthesis*, which commonly deals with the idea of abstracting. What is nowadays usually called *electronics synthesis* does exactly the opposite: it starts from an abstract level and goes to an implementation level.

Within the design cycle, synthesis represents one of the last steps.[5] It follows the specification phase and architectural refinements. The design cycle will be studied in section 1.7 along with other methodological considerations.

Synthesis techniques aim to match the result of architectural choices against a collection of basic elements. An element may be a clearly identified hardware resource or some standard cell for integrated circuits. It can also be a functional resource like a counter, an equality comparator, or a sequencer, which itself is deemed to be implemented by lower-level hardware resources.

---

[5] Two uses of synthesis are not considered to be the last steps of the design, but they are marginal: synthesizing a description to obtain an acceleratable VHDL code or synthesizing a description to quickly have timing or area approximations.

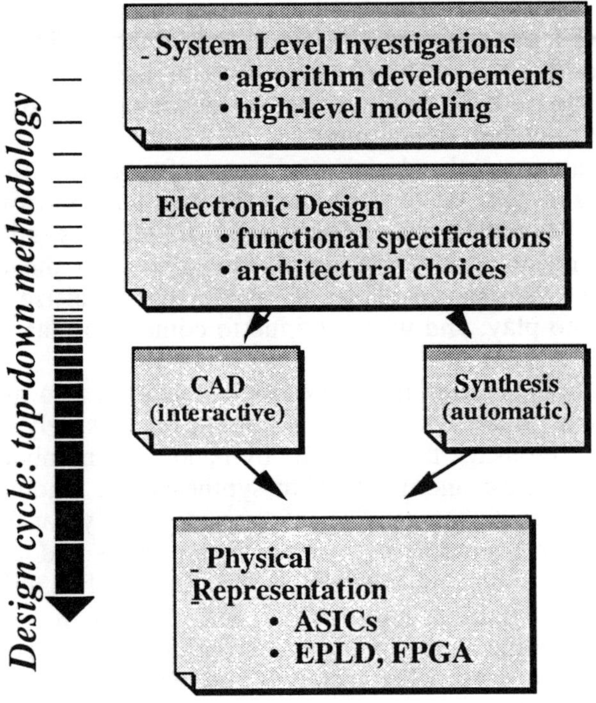


**Figure  1.3.  Design  Cycle**

By the very nature of the targeted resources, two domains are particularly noticeable:

- logic synthesis, whose element library has an explicit hardware equivalent
- architectural synthesis, which is based upon a library of operational resources that are described at a functional level and thus are more abstract

More details on the differences between these domains will be given in section 1.7.


## 1.6. INTERESTS OF SYNTHESIS

From the definition of synthesis given previously, one may infer that its primary advantage is the higher abstraction level of hardware descriptions. The application designer has less and less detailed descriptions to write. For example, he or she is not expected to provide an accurate description of the control signals of a counter explicitly used: only the counting functionality and the constraints on the use of the resource must be indicated. The gain from the abstraction is to reduce and condense descriptions at the start, and, consequently, to ease their debugging.

The more abstract the level, the more functional the specification. Synthesis, in this respect, helps to shorten the path leading from the idea to the implementation. We must however be careful, for there is (still) a long way to go between expressing the functionality and choosing the architecture that optimally fits the constraints. Here is where the expertise and skill of the designer come into play, and will continue to come into play so for some time to come.

Automatically generating hardware implementation also means providing a high-quality result. Since systems are becoming increasingly complex, it is no longer possible to envisage their design "by-hand": the number of objects to be handled is becoming astronomical. With synthesis, the amount of information given by the designer decreases. The information mostly consists of behavioral descriptions and corresponding constraints. Synthesis is undoubtedly leading to more secure designs and should be seen as a quality label on a design cycle.

Since synthesis reduces the size of descriptions, it also allows easier updates, fast debugging, and a wider exploration of architectural choices. In this context, the best tradeoff between cost and performance can more easily be found by the designer.

The large number of lines of hardware descriptions together with the new possibility of parametrization is leading to the creation of (reusable) resource libraries in addition to a better productivity of electronic designs.

The lower average density of layout resulting from the use of synthesis tools compared to hand design is a drawback of such methods. Nevertheless, the evolution of technological processes is leading to a twofold increase in complexity for a given area every two or three years, and therefore offers real possibilities for synthesis.

The increasing complexity of systems to be designed, high-quality requirements, and reduction on the average time to market are other concerns that imply the use of synthesis techniques.


## 1.7. ARCHITECTURAL SYNTHESIS VERSUS LOGIC SYNTHESIS

The synthesis domain consists of two distinct levels: logic synthesis and architectural synthesis. A presentation of these levels to analyze their places in the design cycle, is therefore necessary.

### 1.7.1. Technologies of Logic and Architectural Synthesis

Logic synthesis (also called RTL synthesis), as well as architectural synthesis (also called high-level synthesis), aims to match a behavioral description (the input) with a collection of interconnected elements of a library. These elements have a simple functionality and can be some memory elements (based on latches and D flip-flops), logic operators (and, or, etc.), or arithmetic operators such as adders or multipliers.

Nevertheless, only architectural synthesis accepts more complex operators such as association of elementary operators, sophisticated memories, or ALUs. The main steps allowing the transformation from a text source into a description using these basic elements are the following:

- *Compilation.* The text source is optimized during the analysis phase. Redundancies are eliminated, constants are propagated, and nonaccessible parts of the text source are ignored. These optimization techniques are common to both logic and architectural synthesis and well-known in software compilation.
- *Sharing and allocation of resources.* This is the important point of divergence between logic synthesis and architectural synthesis. Logic synthesis only deals with the sharing of functional resources *used exclusively in time*. This concept needs further explanation. Typically, a description where the output is the conditional result of two additions leads to a sharing of the addition resource: the inputs of the adder are multiplexed (see figure 1.4). Both logic synthesis and architectural synthesis proceed in the same way in this case.

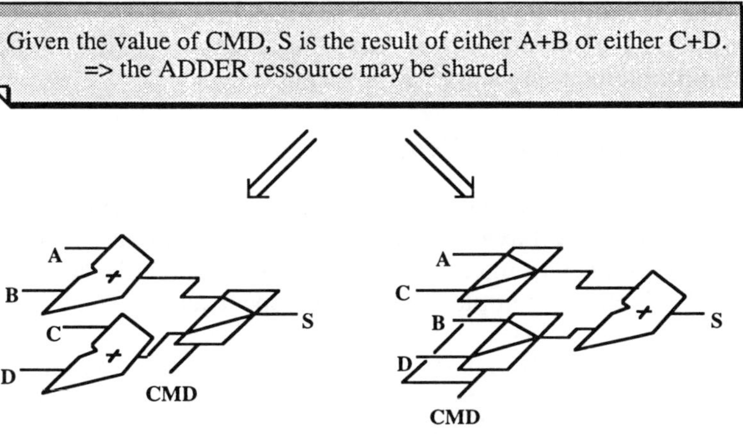


**Figure 1.4. Resource Sharing in Logic Synthesis**

Architectural synthesis goes further: if the timing constraints are respected, these techniques may allocate and reuse the same resource *at different temporal phases*. The real difference with logic synthesis is not only in the interconnection of elements, but also and mainly in the very control of the resource. A significant increase in the abstraction level of the initial behavioral description is induced by this last functionality.

For designers, logic synthesis provides a way to avoid describing required resources. Architectural synthesis will automatically infer not only the resources, but also the control itself, once given the speed constraints as well as timing constraints of the physical elements. The optimum degree of parallelism will therefore be deduced in order to minimize the area or power consumption of the result.

## 1.7.2. Implications on Design Methodology

### 1.7.2.1. New prospects

In the next few years, new synthesis techniques will have more and more impact on system design methodology. The synthesis process by itself is not the source of such a modification. Synthesis is only one of the potential targets of a hardware description.

Modeling ("modelware," that is the act to describe the behavior of a system as a whole), is really changing design methods. The remarkable point is that a single language, VHDL, is now able to cover the entire design cycle from functional specification to low-level structural description.

Methodological elements appear from the ability of VHDL to adapt to all aspects of the problem at any stage of the design.

### 1.7.2.2. Functional specification phase

The goal here is to translate the requirements into a formal description. These requirements may have multiple aspects: documents with texts mixed with schemes, references to international standards (protocol), sets of equations, algorithms, etc. Even the completeness of the specification itself is not always ensured.

The need for a functional description as reference model is therefore obvious. This model will be the essential starting point of the design cycle and will allow, by comparison of simulation results (and perhaps in the future using formal proof), validation of the following steps of the methodology. It is now well-known that such a functional model allows the designer an in-depth understanding of the problem to be solved. Fundamental questions, such as the completeness of data transforms, arise during this phase.

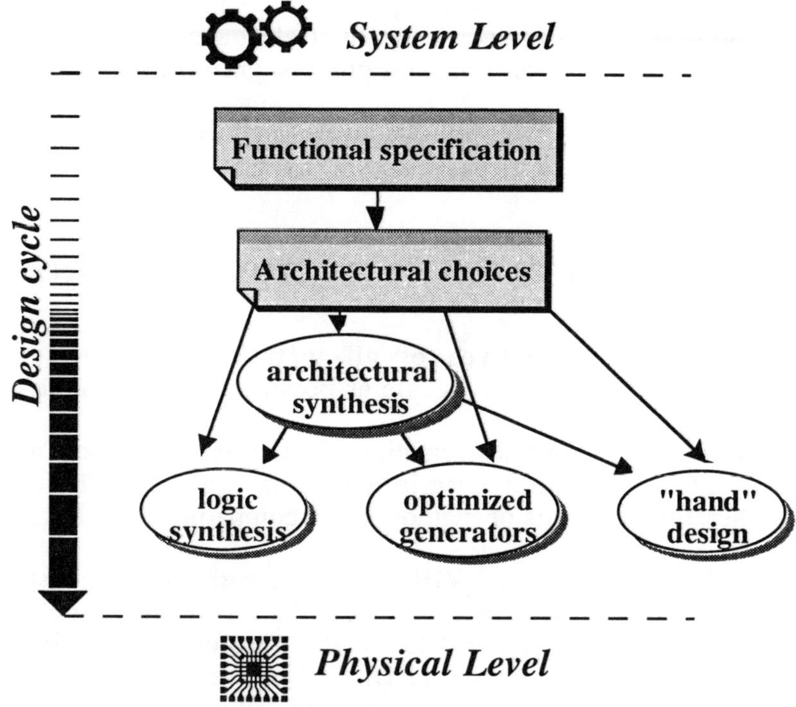



**Figure 1.5. Functional to Physical Level**

### 1.7.2.3. Architectural choice phase

There is no universal tool for this phase. Experience and designer know-how are the keys to success. All the potential targets must be taken into account. Their number is quite large: from software to ASICs, from programmable devices to standard circuits.

Indeed, synthesizable description is also one of these targets. By its very nature, it is fuzzy and abstract. The goal of this book is to help the designer achieve a better understanding of it and therefore use it more efficiently. chapters 3 and 4 study different ways of expressing hardware in VHDL. Therefore, the designer will better understand the real capabilities of today's synthesis tools.

As discussed in the previous section, the synthesis domain is vast. At the highest level, it is possible to express a set of interdependent treatments (data-flow graph) without any specific control. The only requirement is to respect constraints, if any. This is the architectural synthesis domain.

On the other hand, logic synthesis requires the complete description of the control and use of resources in time. Thus, by itself it appears to be a consistent continuation of the architectural synthesis process, once the resource allocation has been performed.

Beyond all doubt, these two synthesis levels (architectural and logic) will in the future merge into one.


## 1.8. CONSISTENCY BETWEEN SIMULATION AND SYNTHESIS

One of the main objectives of all methodology is to make the transformation between two steps as safe as possible. Checks can be provided for this.

Synthesis consists of the transformation of an initial description, as abstract as possible, into a structural description using well-identified hardware resources. Therefore, in terms of methodology, comparing both descriptions is essential.

The power of description of VHDL is able, without any constraints, to support the two levels of description. Indeed, there is no real problem at this stage, but much more trouble is related to the transformation technology. In other words, the way of interpreting VHDL text sources for synthesis purposes (the semantics for synthesis) is fundamental and critical, and it determines the quality of the result.

It is true that the design of VHDL has only been achieved in terms of simulation semantics but this book wishes to proclaim loud and clear that an unambiguous synthesis semantics, respecting the simulation semantics, may be defined. It is time now to go into more detail.

# 2. VHDL  CONCEPTS

*In this chapter, the grammar of some VHDL constructs is given. This is a correct but simplified form that allows for usual descriptions with a view to synthesis. More complete forms are given in appendix 8.1 and, of course, in appendix A of the LRM (Language Reference Manual).*

## 2.1. PHILOSOPHY OF THE LANGUAGE

### 2.1.1. Generality

VHDL allows for the description of hardware behavior from system to gate levels. Figure 2.1 shows the different possible description levels. A rough characterization of these levels can be provided. A certain number of expression needs, covered by the language, are associated with each level.

*The system level* focuses on the description of the functionalities of the system (what it does) and tries to avoid its implementation description (how it is constituted). The notion of time is essentially a notion of causality: one action implies another. A constant concern is to forget useless details, which would imply architectural choices too early in the design methodology. Too detailed a system description is a drawback for it restricts further architectural choices or implies a given technology. Therefore, hiding the information structure is desirable and the notion of concurrency may not be necessary at this phase.

To fit this level of description, the language has to offer large degrees of abstraction, a powerful algorithmic, wide capabilities for merging different description levels, an easy expression of causality, and also the possibility of introducing nondeterminism, which may be an interesting feature.

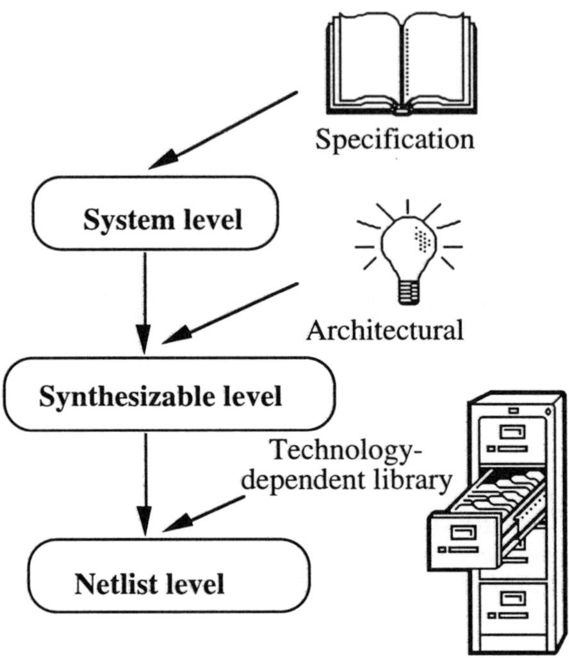


**Figure 2.1. Different Levels of Description**

To date, this level of description has not been synthesizable: no explicit architecture is described and no tool on the market offers a real and efficient architectural synthesis (except for some specific target architectures). As a result, this level will not be studied in detail in this book. Nevertheless, it could be useful to remember that a VHDL description does not have to be entirely synthesizable: the testbench is a nonnegligible part of this description and can be described at system level.

*The synthesizable level* is the potential input for synthesis tools. Here, some implementation choices have already been made:
- an architecture, or at least an architecture family is targeted and is implied by the code structure
- the widths of datapaths are known
- time can be expressed in terms of clock or sequentiality (one statement is executed after the previous one)

The language must allow a description of the model at this level with a sufficient level of abstraction towards the physical level. Clocks, sequentiality, dataflows, and combinational parts have to be easily expressed. A large degree of parametrization is also required.

*The netlist level* is the potential output of synthesis tools. It is a structural view appearing as a collection of model instantiations. This kind of description involves the existence of model libraries. The notion of time is often present in the description of these models, from the notion of propagation delay through a gate to very sophisticated delays (using slopes, temperature, etc.). These delays are either provided by the library and are therefore only indicative, or are relevant to a given technology and therefore more accurate. They can even be deduced from the netlist using a backannotation mechanism (probably outside the VHDL world).

At this step, the language has to offer an optimal flexibility in terms of timing configuration or technology. These two aspects are most often linked.

### 2.1.2. Time

The notion of time, which is carefully described in the LRM, is only related to simulation. This time is discrete: only events have a date. Indeed, the notion of time does not exist between two of these dates. The simulator is event-driven: it skips from one event to the following one without exploring what happens in between.

No synthesis semantics is defined in the VHDL LRM. Therefore, it is not possible to directly express implementation timing constraints using the language. The only existing notion is the delay: "this output takes this value after this exact delay."

Moreover, no MIN/MAX simulation mechanism is included in the language.

### 2.1.3. Modularity

A VHDL description is never monolithic; modularity is everywhere. The first structuring level is the *design unit.*

When compiling a VHDL source file, this file notion does not exist after compilation: each contained design unit once successfully analyzed is independently stored within a VHDL *library.*

The only structuration of the original file which exists after compilation in the VHDL world is the notion of design unit. As shown in figure 2.2, a VHDL source file is seen as a collection of design units. A design unit cannot be split between multiple source files, and there is no implicit link between design units written in the same source file.

The VHDL basic design cycle consists in first compiling *source files*, and then recompiling or executing (simulating, synthesizing, or proving) *design units*.

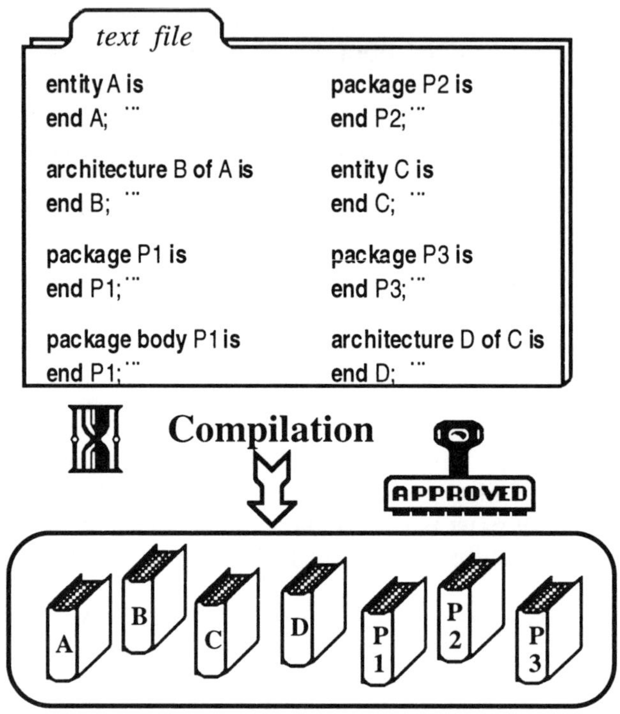


**A set of design units in library WORK**

**Figure  2.2.  Compiling  Source  File  into  Design  Units**

There are five kinds of design units. They can be roughly split into two parts: those that describe the hardware hierarchy (i.e., the structure of the model) and those that are more software oriented. This separation is shown in figure 2.3. An orthogonal classification shows that three design units are "primary," and two are "secondary." A primary unit defines the interface with the external world. A secondary unit is the implementation (internal view) of its primary.

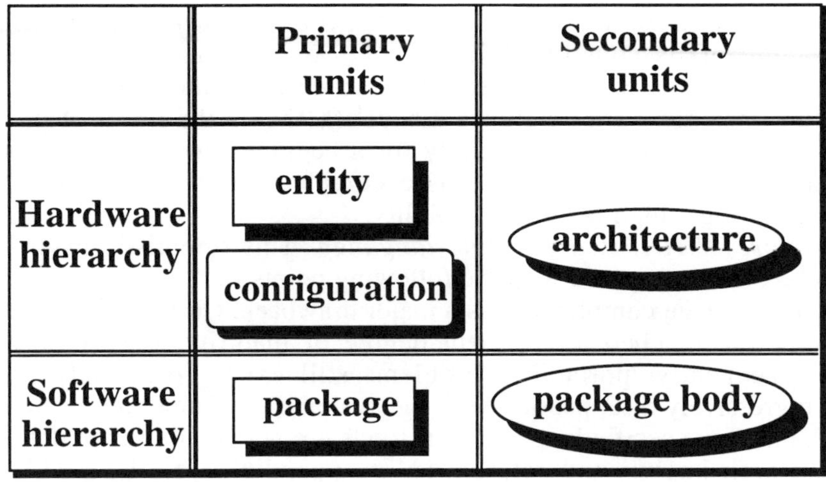


**Figure 2.3. Five Kinds of Design Units**

A varying number of design units, possibly of different kinds, constitute a library. Each design unit is self-compilable but may use objects of other design units, possibly stored in other libraries.

```
library RESOURCE;        -- allow access to library RESOURCE
use RESOURCE.P.all;      -- allow access to the objects of package P by their simple name
entity A is
    …
    -- use of some (exported) object, for example a type, of package P
    …
end A;
```

In the above example, design unit A uses a data type that is defined in a package P stored in a library called RESOURCE. These dependencies (unit A is said to be "dependent" on unit P, because "using" it) imply a compilation partial order: package P has to be compiled before the first compilation of entity A, even if they are not in the same source file. Likewise any change in the external view (package declaration) of P involves a recompilation of entity A.


## 2.1.4. Portability

Portability was one of the main guidelines during the VHDL language design phase. The widespread use of this language is mainly due to the fact that it is a standard. A standard is the only way for users to be free of the potential precariousness of a proprietary language.

For a language to be portable, its syntax (its constructs) and its semantics (the meaning of its constructs) have to be as clear and nonambiguous as possible. Meticulous work has been done to describe the VHDL syntax and simulation semantics. The reference manual (LRM) is the result of this work. Indeed, although its clarity, especially for a beginner, may be debatable, this document offers a real interest: it is a complete reference for VHDL tool builders ensuring a high level of portability.

*Therefore, VHDL is a portable language.* This optimistic assertion has nevertheless to be tempered with the following points:

- The language complexity has a major drawback. Certain implementations have not reached a sufficient degree of maturity to fit the standard. Consequently, portability problems still exist. Fortunately, they are progressively disappearing with the availability of new tool versions.
- Some very specific language constructs are potentially nonportable. They are well-known and referenced in appendix C of the LRM. As an example, computations with real types can lead to different results due to computer accuracy.
- Unfortunately for designers using synthesis tools, only the simulation semantics of the language is defined in the LRM. This lack implies the necessary definition of modeling practices for synthesis. They are detailed in the following chapters. Note that we are in a consensus domain, sometimes it is a widely accepted consensus but never a real standard practice.

Nevertheless, using such practices, designers can potentially use a large range of simulation and synthesis tools.


## 2.2. HARDWARE HIERARCHY


### 2.2.1. Entity/Architecture


#### 2.2.1.1 Entity

As shown in figure 2.4, an entity is the external view of a model: *ports* (inputs/outputs) and parameters (named *generics*), as well as static checks on parameter values (such as range or minimal values verification) and dynamic checks on ports (such as set-up or hold time verifications) are described in this design unit.

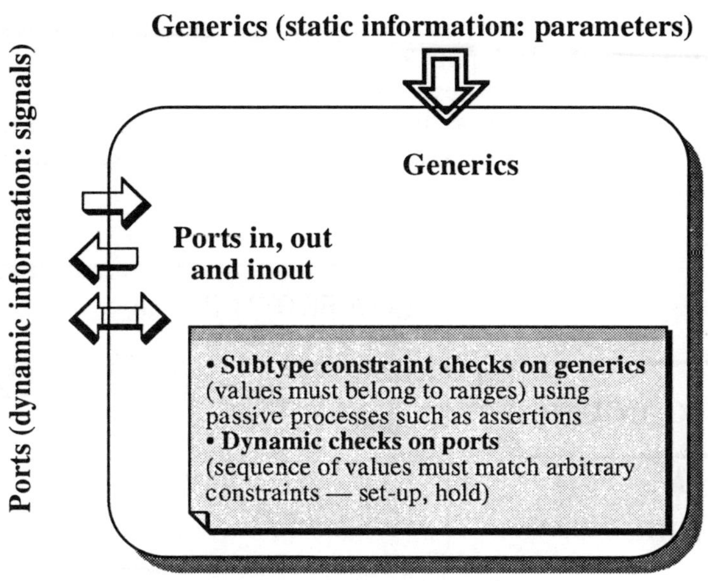


Figure  2.4. Concept  of  Entity

Its syntax is

```
entity identifier is
            [ generic ( generic_list ) ] [ port ( port_list ) ; ]
            entity_declarative_part
[begin
            {passive_concurrent_statement}]
end [ entity ] [ entity_simple_name ] ;
```

As illustrated in figures 2.5 and 2.6, this external view does not depend on the description level.

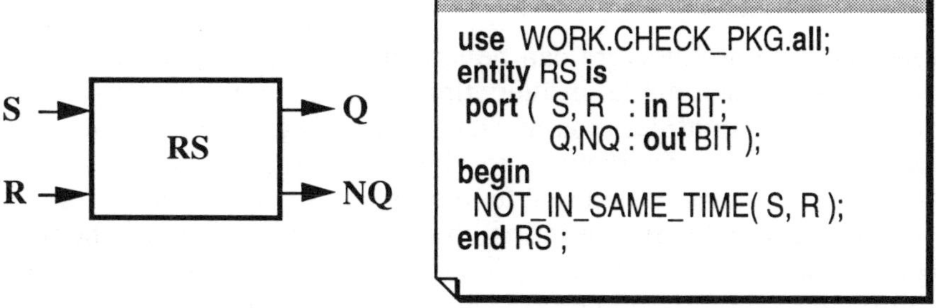


Figure  2.5. Circuit-Level  Entity  —  RS  Latch

Figure 2.5 shows an entity modeling the external view of a RS latch. The ports and a dynamic check (inputs R and S should not change at the same time) are described. This is a circuit-level description.

Another level of description, the system level, is presented in figure 2.6. It shows the high-level description of a coder with ports, parameters, and decoding law and mode.

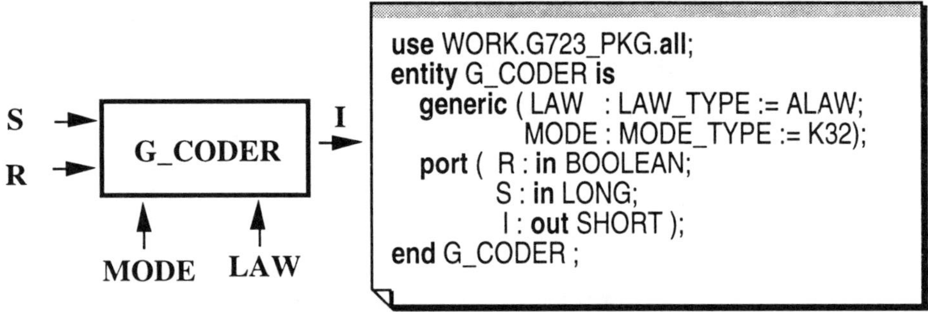


**Figure 2.6. System Level Entity — Coder**

### 2.2.1.2 Architecture

The structure of a model, its behavior, or any mixture of both structure and behavior are described within its architecture. The syntax of such a design unit is

```
architecture identifier of entity_name is
     architecture_declarative_part
begin
     all_concurrent_statements
end [ architecture ] [ architecture_simple_name ] ;
```

More than one architecture can be associated with a given entity but only one is selected for each model instantiation at simulation time. This selection is the main goal of the configuration mechanism that is discussed below.

Allowing multiple architectures for the same entity is of great interest. For example, after synthesis, it is possible to compare the two architectures (before and after synthesis) by instantiating them (taking a copy of them) in two different instances. Certain synthesis tools even propose different architectures as output of the synthesis process. Each one is optimized for a given purpose: accurate simulation, quick simulation, or hardware acceleration.

An architecture is implicitly dependent on its associated entity: all objects defined in the entity are known within the architecture. Therefore, as shown in figure 2.7, ports (here NQ and Q) are seen as signals and can be assigned within the architecture.

The architecture body description consists of a set of concurrent statements. They will be presented in section 2.6.

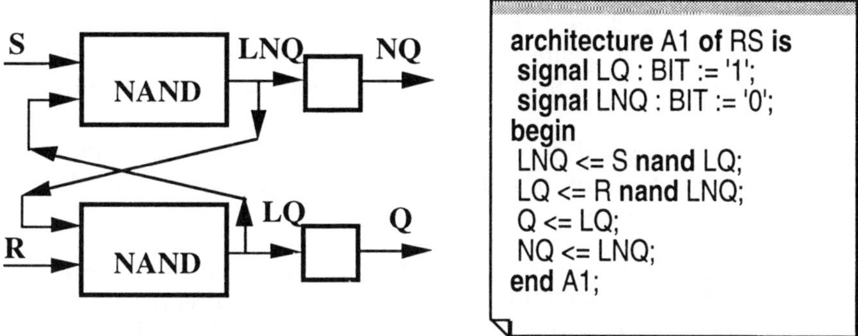


**Figure  2.7. Structural Architecture at Gate-Level — Latch**

```
architecture B1 of G_CODER is
begin
        COD : process
                variable CONTEXT : CTX_CODER_TYPE;
                variable SORTIE : SHORT;
        begin
                wait on R'TRANSACTION, S'TRANSACTION;
                CODER (R, LAW, MODE, CONTEXT, S, SORTIE);
                I <= SORTIE;
        end process COD;
end B1;
```

**Figure  2.8. Behavioral Architecture at System-Level — Coder**

### 2.2.2. Direct Instantiation of Entity

Instantiating means "taking a copy" of a model. When a circuit can be described using three nand gates, the same "nand" model is used and instantiated three times: for example, instances NAND1, NAND2, and NAND3 are created.

The simplest VHDL mechanism for creating an instance is to instantiate an entity/architecture pair. *This mechanism was not allowed in the VHDL'87 standard; therefore, it is not yet implemented by all VHDL tools.* Easy to use, its main drawback is to create a permanent link between instances and pairs entity/architecture. Figure 2.9 illustrates this mechanism by instantiating three circuits on a board.

From a methodological point of view, this mechanism must be avoided as often as possible. The notion of component instantiation, introduced in the next section, should be preferable.

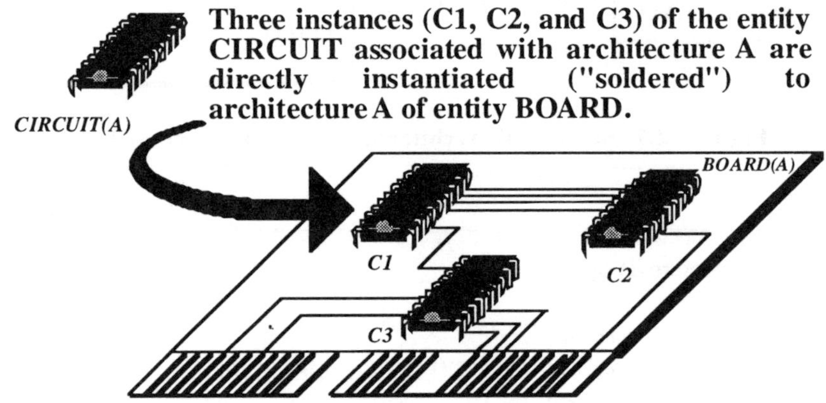


Figure 2.9. Direct Instantiation

Its syntax is

```
label : entity entity_name [ ( architecture_identifier ) ]
        [ generic map ( generic_association_list ) ] [ port map ( port_association_list ) ] ;
```

### 2.2.3. Notion of Component

This is a powerful VHDL notion. Unfortunately, the notion of component is not natural for beginners.

In a first approach, we have seen that a model may be instantiated a certain number of times: a nand gate model can be instantiated three times and then provides the gates P1, P2, and P3. Therefore, a single model, i.e., a single description, generates different instantiations, each of them having its own behavior.

To be more general, VHDL offers a general and flexible mechanism: the instantiation is applied to the idea we have of a model, and not as in the case of direct instantiation to the model itself.

It is therefore possible, either to directly instantiate a pair entity/architecture (as seen in paragraph 2.2.2), or to instantiate an intermediate object called *component*. Indeed, this notion of component represents the external view of a desired model. This desire may be quite different from the actual external view of the model we will finally use, and adaptation mechanisms are provided.

There is no behavior attached to the notion of component, but it has one great advantage: it allows compilation. So, many static checks may be performed even if the entity/architecture pair that will finally be used is not known (or does not even exist yet).

This mechanism allows for a top-down methodology with real decoupling between the component library (what is desired) and the model library (what we have).

Using the notion of component implies three fundamental operations: declaration, instantiation, and configuration of the component. Fortunately, and especially in the logic synthesis domain, many tools generate this source code automatically.

### 2.2.4. Component Declaration

When the designer describes both the component (his or her requirement) and the entity (what he or she will actually use), the component declaration and entity declaration tend to be very similar. This fact is often pointed out by VHDL opponents to show the verbosity of the language.

```
architecture A of TEST is

        component INVERTER        -- component declaration: what we wish
            port (INPUT : in BIT; OUTPUT : out BIT);
        end component;

begin
        ...
end A;

entity INV is      -- entity declaration: what we have
            port (INPUT : in BIT; OUTPUT : out BIT);
end INV;
```

Nevertheless, considering the component declaration as a simple redundancy of the entity declaration is an error. When using already existing libraries or design units written by somebody else, the power of the component notion appears obvious: adaptation is possible.

```
architecture A of TEST is
        component INVERTER         -- component declaration: what we wish
            port (INPUT : in BIT; OUTPUT : out BIT);
        end component;
begin
        ...
end A;

entity INV is      -- entity declaration: what we have
            generic (TP : TIME);
            port (OUTP : out BIT; INP : in BIT);
end INV;
```

In this case, certain adaptations have to be performed: the model has an extra parameter (TP), different port names, and order. The component configuration allows these adaptations to be performed.

## 2.2.5. Component Instantiation

Roughly speaking, instantiating a component consists of "soldering" on the board the socket where the circuit will be plugged. This second operation, plugging, is the purpose of the component configuration. The name of the component as well as the way component ports are associated with signals (the wires on the board) are given when instantiating. Figure 2.10 illustrates this operation.

**Instance C1 of component COMP is instantiated in architecture A of entity BOARD**

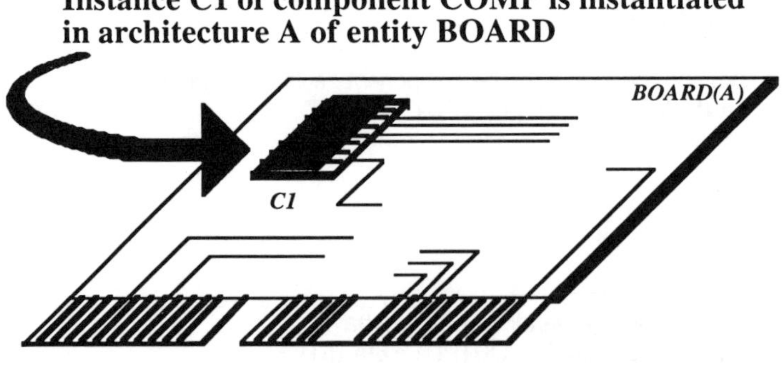


**Figure  2.10.          Notion  of  Component  Instantiation**

The syntax of this statement is

```
label : [ component ] component_name
        [ generic map ( generic_association_list ) ] [ port map ( port_association_list ) ] ;
```

The following lines show the VHDL source code of a collection of instantiations (C1, C2, and C3):

```
architecture A of TEST is
        component INVERTER -- component declaration: what we wish
            port (INPUT : in BIT; OUTPUT : out BIT);
        end component;
        signal A, B, C, D : BIT;
begin
        C1 : INVERTER port map (A, B);
        C2 : INVERTER port map (B, C);
        C3 : INVERTER port map (C, D);
end A;
```

At this level of description, no information is given on the way the component will be configured (i.e., what will be the associated entity/architecture pair). In practice, this pair possibly does not even exist at this stage. If a separate design unit, called a configuration unit, is used for the component configurations, the previous code can be compiled as it is but not executed.


## 2.2.6. Component  Configuration

Associating entity/architecture pairs to component instances is the goal of this operation. It can be performed in the architecture where components are instantiated or within a specific design unit: the configuration unit.

In both cases, it allows an adaptation of the component to the model (the pair entity/architecture). This adaptation may consist in changing the name of ports or parameters, also called generics, modifying their order, and even making some of them disappear. The syntax of this operation, when declared in the declarative part of the architecture is

```
for instantiation_list : component_name
                [ use entity entity_name [ ( architecture_identifier) ] ]
                [ generic map ( generic_association_list ) ] [ port map ( port_association_list ) ] ;
```

Continuing our analogy, this operation can be seen as plugging a circuit into a socket (cf. figure 2.11). Each socket corresponds to a component instantiation. Adapting the socket to the circuit is possible during this operation.

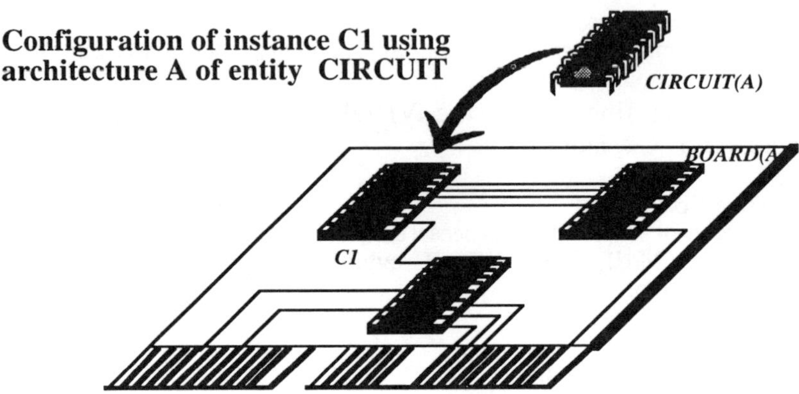


**Figure  2.11.**         **Configuration  of  a  Component  Instantiation**

In the following source code, instances C1, C2, and C3 are configured to
the previous entity INV and its architecture B. The adaptations of parameters
and ports are performed through **map** clauses.

```
architecture A of TEST is

        component INVERTER -- component declaration: what we wish
            port (INPUT : in BIT; OUTPUT : out BIT);
        end component;

        for C1, C2, C3 : INVERTER use entity WORK.INV (B)
                    generic map (10 ns) port map (INP => INPUT, OUTP => OUTPUT);
        signal A, B, C, D : BIT;

begin

        C1 : INVERTER port map (A, B);
        C2 : INVERTER port map (B, C);
        C3 : INVERTER port map (C, D);

end A;
```

The flexibility provided by the notion of component is very powerful.
Selecting an entity/architecture pair is possible very late in the design cycle (just
before simulation or synthesis) and switching from one library to another to
change one model into another (cf. figure 2.12) is a straightforward operation.

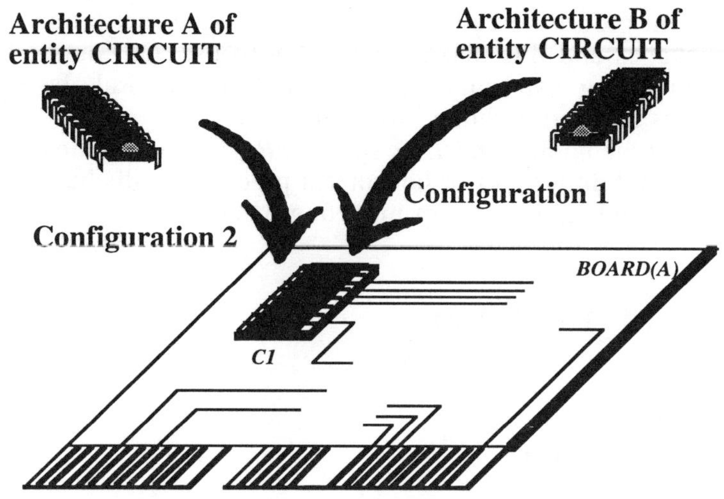


**Figure  2.12.        Flexibility  Carried  Out  by  the  Configuration**


## 2.2.7. Concurrent  Procedure  Call

When called from a sequential context (from within a process or a subprogram), a procedure has the semantics of a subprogram, well-known in programming languages. However, calling it from a concurrent context (from within an architecture or a block) gives it another dimension.

In this case, if certain of its formal parameters of mode **in** are declared to be of class **signal**, the procedure call will automatically be performed each time an event occurs on at least one of these signals.

There are two main applications for concurrent procedures:

- When passive, without any parameter of mode **out** or **inout** of class signal and thus without signal assignment, they allow dynamic and automatic checks of causality or timing constraints (set-up, hold, etc.).
- When active, they may be used to assign signals (mode **out** or **inout** parameters of class **signal**) conditionally to events occurring on other signals (mode **in** or **inout** parameters of class **signal**). In fact, they offer an alternative to instantiation for modeling hardware components. This is the reason for considering them as possible parts of the hardware hierarchy.

Since this second possibility is close to the notion of instantiation, its interest has to be clearly defined. All the flexibility previously discussed concerning the notion of component (and configuration) is obviously lost when using concurrent procedure calls. Nevertheless, concurrent procedures may be stored in packages and this makes them easy to use. Of course, this kind of hardware structuring can only be applied to behavioral descriptions: no structural part is allowed within a procedure.

Chapter 6 of the book *"VHDL Designer's Reference"* studies in more detail the advantages and drawbacks of concurrent procedure calls but the decision to use it or not is mainly a question of modeling practice.


## 2.3. SOFTWARE HIERARCHY


### 2.3.1. Writing Software

Talking software in a language devoted to hardware can seem inconsistent. In practice, software parts are often useful in a hardware description. These parts mainly target two different needs:
  • writing behavioral descriptions by structuring sequential statements
  • developing tools (trace, pattern generator, etc.)

Structuring these parts in subprograms and packages is an efficient way of facilitating the reuse of such parts.


### 2.3.2. Packages

As shown in figure 2.13, there is an obvious analogy between the notions of package declaration/package body and those of entity declaration/architecture body. The VHDL software hierarchy is based on packages in the same way hardware hierarchy relies on entities/architectures pairs.

Objects of the language (subprograms, types, constants, signals, etc.), which are declared in the package declaration design unit, are said to be exported by the package. Any design unit referencing this package (with a **use** clause) has access to these objects and may use them. The general syntax of this design unit is

```
package name_of_the_package is
        package_declarative_part
end [ package ] [ name_of_the_package ] ;
```

Information relevant to the way exported subprograms are built (algorithms, local objects) are within the package body whose syntax is

```
package body name_of_the_package is
            package_body_declarative_part
end [ package body ] [ name_of_the_package ] ;
```

We are very close to the notion of external/internal views of hardware hierarchy: entity and architecture.

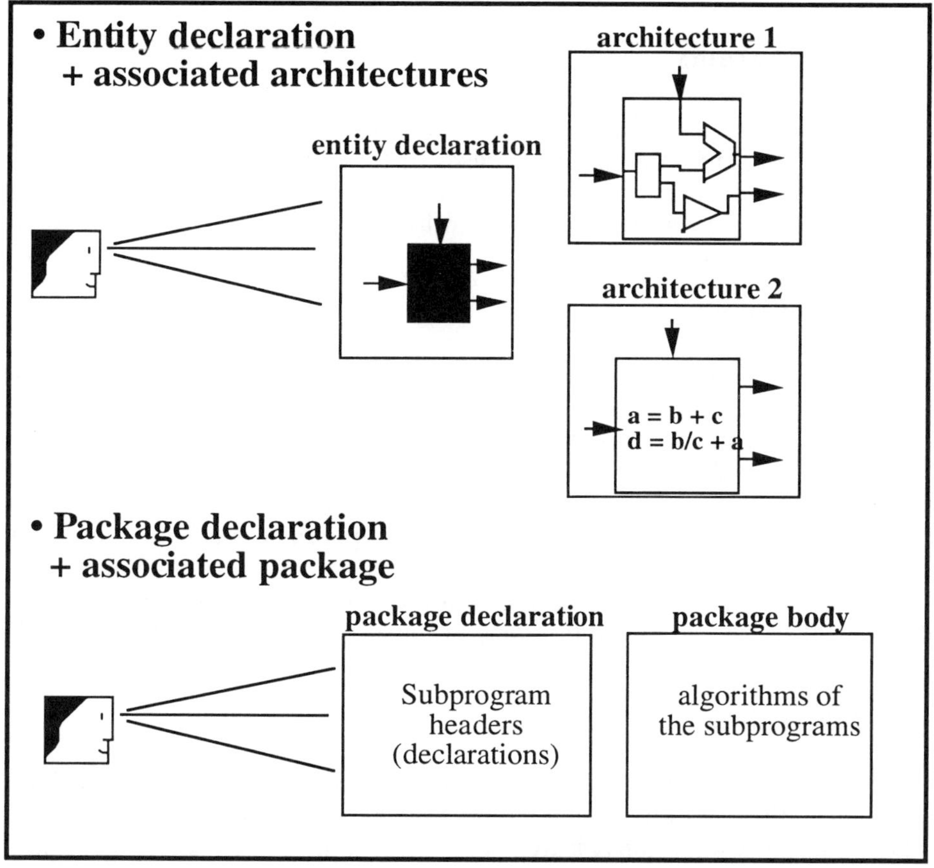


**Figure 2.13.     Duality Between Hardware/Software Hierarchies**

Packages are an efficient way of standardizing a VHDL environment without adding anything to the LRM. The type STD_LOGIC, standard nine-state logic value, which is very popular in synthesis, is a good example. It is entirely defined (the type itself and a collection of subprograms for using it) in a package. The header of the subprograms, defining their profile (function or procedure, their names, and parameters) is in the package declaration but corresponding algorithms remain in the package body.

When correctly documented, the knowledge of the package declaration is sufficient to call the subprograms.

Two standard packages, STANDARD and TEXTIO, are defined within the LRM. They will be described further in section 2.9. A certain number of packages, even if not defined within the LRM, are nevertheless part of the standard environment for synthesis and are detailed in section 2.9.


### 2.3.3. Subprograms

Functions and procedures are the two kinds of subprograms. Operators (logic, arithmetic, etc.) are particular cases of functions.

Functions and operators have a various number of input parameters but only return a single value. As in the following example, they are called in expressions.

    B := MY_FUNCTION(5, A) + C;

In *VHDL'87*, functions cannot have side effects. This means that they only access locally declared objects or objects passed as parameters. Consequently, two calls to the same function with the same parameter values return the same value. In *VHDL'92*, this notion of pure function becomes optional. Impure functions, using side effects, are legal. For synthesis purpose, this difference between pure and impure function is of some importance: pure function implies combinational hardware and impure function may involve memorization.

Procedures have potentially any number of input or output parameters. Bidirectional parameters (input/output) are also possible. A procedure call is a statement by itself.

    MY_PROCEDURE(A, B, C);

Subprograms may be declared in different declarative parts but, if defined in packages, they acquire much more power. A subprogram *exported* by a package may be used in any design unit referencing the package with the **use** clause. To export a subprogram, its header (the external view of the subprogram) has to be declared within the package declaration. Here is the syntax for declaring procedure and function headers:

**procedure** name_of_procedure [ ( interface_element[6] { ; interface_element } ) ] ;

[ **pure** | **impure** ] **function** name_of_function
             [ ( interface_element { ; interface_element } ) ] **return** type_of_returned_value ;

---

[6] Syntax of interface element is given in appendix 8.1.

In the case of exported subprograms, the subprogram body (the internal view describing the algorithm) is written in the corresponding package body. Its syntax is the following:

```
procedure name_of_procedure [ ( interface_element { ; interface_element } ) ] is
          subprogram_declarative_part
begin
          all_sequential_statements
end [ procedure ] [ designator ] ;

[ pure | impure ]  function designator
    [ ( interface_element { ; interface_element } ) ] return type_of_returned_value is
          subprogram_declarative_part
begin
          all_sequential_statements_(except_wait_statement)
end [ function ] [ designator ] ;
```

Therefore, as for packages, subprograms are divided into the external part (header) and the internal part (body). The header describes the parameter profile (name, number, type, direction) and is naturally placed within the package declaration. Subprogram bodies are the main constituent of the package body (as shown in figure 2.14).

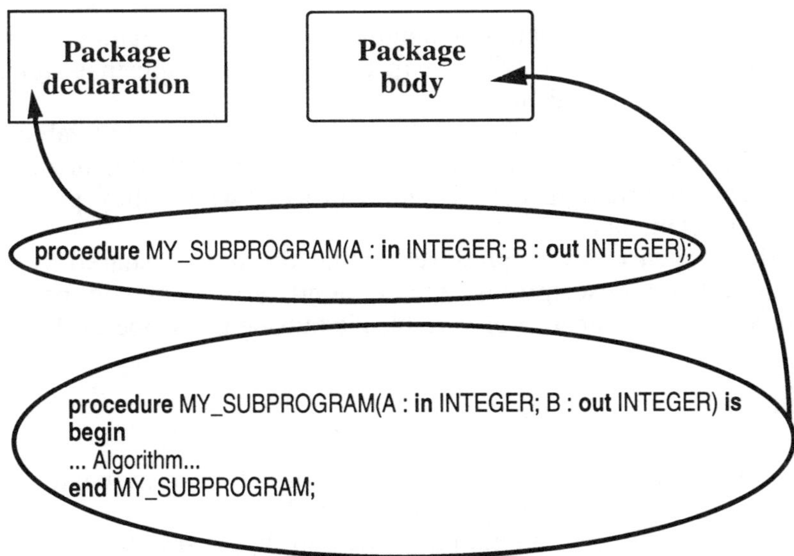


**Figure  2.14.**          **Subprogram  Header  and  Subprogram  Body**

Overloading is an important characteristic of VHDL subprograms: two (or more) subprograms having the same name are allowed if their profiles are different: i.e., if the number, order, or type of parameters differ. Overloading is widely used in VHDL. For instance, standard package STD_LOGIC offers a set of logical operators able to handle types BIT as well as BOOLEAN; operators for each of these two types are simply overloaded.

## 2.4. OBJECTS OF THE LANGUAGE

### 2.4.1. Constants

Constants are well-known objects in programming languages. Once their values are defined during initialization, these values remain unchanged throughout the execution. However, when declared within a subprogram, constants are recomputed each time the subprogram is called but remain constant during the execution of the subprogram.

### 2.4.2. Variables

Variables, as in programming languages have their values immediately updated when assigned. We will see in the next paragraph that signals offer another assignment mechanism.

Two kinds of variables exist: regular variables and shared variables.
- Regular variables can be declared either in subprograms or in processes and are important in the synthesis process where they possibly infer memory elements.
- VHDL concurrency implies restricting the use of shared variables to system level descriptions, where nondeterminism can be accepted. Therefore, this category of variables is outside the scope of this book.

### 2.4.3. Signal

On PCB or within integrated circuits, information is propagated through wires. In VHDL, these wires are called "signals."

Signals exist from the beginning to the end of the simulation process. No creation of new signal or physical disconnection of signals are allowed during simulation. In the real world, when the output of a ROM is three-stated, this output does not physically disappear. Its value (its impedance) only implies a theoretical disconnection. VHDL mechanisms are available for emulating this notion.

The notion of clock, as a specific signal, does not exist in VHDL. As illustrated in chapters 3 and 4, this fact implies clock recognition problems for synthesis tools when parsing a VHDL description.

It is important to note that information exchanged between components only transits through signals. The global VHDL netlist, the entire set of components interconnected with signals, exists throughout the simulation process. Even if signals seem to have the hardware meaning of wires, for optimization purposes, some of them may be merged during the synthesis process.

Signals are not containers as variables are. As seen previously, signals are permanent objects and have fixed links with other signals. Signals have a past, accessible using attributes (kind of functions), a current value, and a future consisting of a set of projected waveforms managed by the simulator kernel.

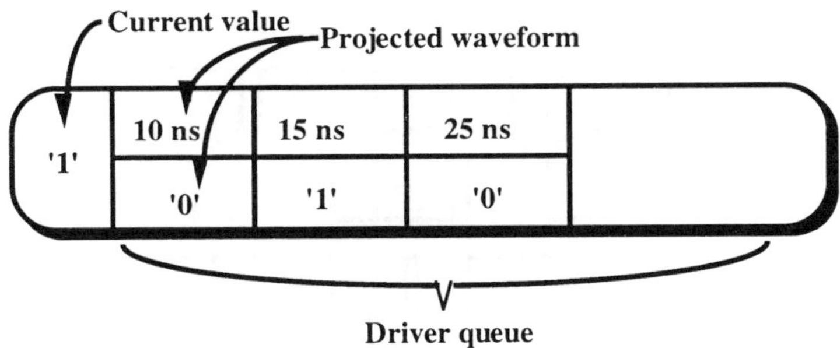


**Figure  2.15.**        **Usual Representation of a Driver Signal**

One of the syntactic forms of signal assignment is the following:

```
[ label : ] target <= [ transport | inertial ] waveform ;
```

The main characteristic of a signal comes from the semantics of this assignment. Using symbol "<=", a signal assignment may involve the result of an expression, possibly using other signal current values.

But, the signal assignment

```
A <= expression_B;
```

does not imply that the current value of signal A is immediately updated. Indeed, the value of expression_B is stored in the driver of A as a projected value of A. It will only become the current value of A when encountering a synchronisation statement (**wait** statement). The whole concept of concurrency relies on this signal assignment mechanism.

Furthermore, this mechanism may use delays:

```
A <= expression_B after 15 ns;
```

Even if such an **after** clause has no meaning for synthesis tools and is (in the best case) ignored, it contains very accurate simulation semantics. This meaning has to be known by the designer: the testbench often consists of stimulus description involving such clauses. Figure 2.16 illustrates the results of a signal assignment with inertial delay (signal assignment mode taken as default) and transport mode. Inertial mode signal assignment is often used in synthesizable descriptions: it allows an accurate description of gate behavior by delaying signals and filtering small pulses (glitches). The behavior of the transport assignment statement is given here for information purposes, this ideal transmission mode being mainly used for transmission line modeling.

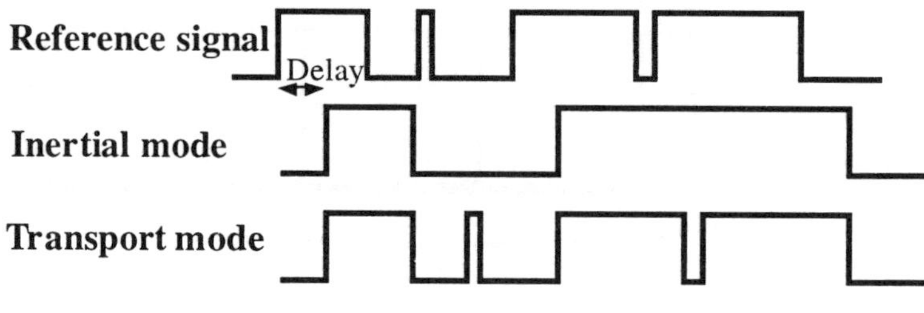


**Figure 2.16.        Inertial and Transport Modes**

## 2.4.4. Files

The status of files was not very clear in *VHDL'87*: type or object class. The *VHDL'92* version definitely considers them as a specific class of objects. It is impossible to assign them, but of course read and write operations may be performed on them. They have no meaning in terms of synthesis and therefore are reserved to nonsynthesizable parts of the description: testbench or trace, for example. The only noteworthy exception is their possible use in constant computations.

## 2.5. INFORMATION  REPRESENTATION

### 2.5.1. Types  of  Objects

What is the structure of information in a variable, a constant value, or a signal? At a physical level, it can appear as voltage. Since VHDL is not aimed at analog description, '0' and '1' values can be considered. This is a first approach of the notion of type: type BIT. It is possible to go further in our example and to define 'Z' as a tristate value, 'L' and 'H' as respectively weak '0' and weak '1', etc. A more complete type, involving nine different logic values, has been especially defined for synthesis purposes and will be further described in section 6.2.

Structuring information is one way to provide more abstraction. For example, handling "words" is common when dealing with bit vectors. This capability is offered in VHDL, and therefore a single signal may be used to represent a complete bus. Of course, more abstract types can be imagined. For example, a signal may model a microstatement with its address part, its operative part, and data part. Records are often used for this purpose.

A lot of the description power of VHDL comes from this capability for the designer to define and use his or her own types. Furthermore, a large number of compilation checks ensure the consistency between values and operations performed on them.

There are four families of VHDL types:
- Scalar types, whose value is made up of a single element.
- Composite types, whose value involves several elements.
- Access types, which are the "pointers," well-known in programming languages.
- File types, which are only types for file objects.

The last two families will not be described in detail in this book. Access types are mainly useful at the system level and have no direct relationship with hardware description. Files have obvious applications in storing stimuli or traces but are not used in synthesis.

### 2.5.2. Scalar  Types

There are four kinds of scalar types:

- Enumerated types, whose possible values are explicitly listed:

```
type SEVERITY_LEVEL is (NOTE, WARNING, ERROR, FAILURE);
```

• Integers, whose range of values may be specified:

```
type MEMORY_SIZE is range 1 to 2048;
```

Defining a constraint on the range of such types implies that checks will be performed during simulation. Any constraint violation will generate an error. Furthermore, such an indication allows synthesis tools to compute the number of bits needed for representing objects of that type.

• Real types, which may also have an explicit range:

```
type ACQUISITION is range 1.0 to 9.0;
```

• Physical types, which can be considered as integers associated with a physical unit that allows easy notation and dimension analysis checks:

```
type TIME
    is range -9_223_372_036_854_775_808 to 9_223_372_036_854_775_807
    -- Implementation-defined range
    units fs;
        ps = 1000 fs;  ns = 1000 ps;  us = 1000 ns;  ms = 1000 us;
        sec = 1000 ms; min = 60 sec; hr = 60 min;
    end units;
```

Physical types, are not recognized (and thus not used) in synthesis.


## 2.5.3. Composite  Types

These types are useful for grouping conceptually bound information. The first of them, arrays, are useful for structuring elements in a single type. For example, a bus can be seen as an array of bits.

```
type WORD is array (0 to 63) of BIT;    -- 64-bit bus
```

For synthesis purposes, this information has nevertheless to be completed with conventions in order to give more detail on the structure: Is there a sign bit? Where is it? Where is the most significant bit?

An array element is designated using an index. Given an object A (constant, variable or signal) of type WORD, A(5) designates the sixth element of A (if the array index of A begins at 0).

Records are the other kind of composite types. They are less widely used for synthesis purposes but can sometimes be very useful.

```
type INFORMATION is record
      VALUE : BIT;
      CONTEXT : INTEGER;
end record;
```

Accessing a record element is possible using its name. Given B, an object of type INFORMATION, B.CONTEXT designates the integer element of the value of B.


## 2.5.4. Object Classes

As previously seen, four different classes of objects exist in VHDL: constants, variables, signals, and files. The notions of classes and types are conceptually different. *Type* describes the structure of the information whereas *class* is relevant to its nature.

A constant may be based on a record type; variables, signals, and even files may also. This type indicates that the value of these objects is made up of different elements. On the other hand, the information carried by these objects may be of a different nature: static (for a constant), local and dynamic (in the case of a regular (i.e., not shared) variable) or global and dynamic (for a signal or a shared variable).


## 2.6. CONCURRENCY


## 2.6.1. Informal Definition

Concurrency is an important and natural concept in hardware description. A microprocessor computing at the same time a memory is providing data is an example of two hardware devices working concurrently. Concurrency does not only concern a high level of abstraction, a simple latch as shown in figure 2.17 also illustrates this concept: both gates are "computing" together.

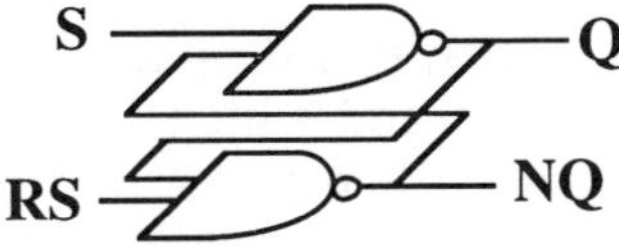


**Figure   2.17.        Latch — Two "Concurrent" Gates**

One possible VHDL description of this netlist may be the following two statements, supposed to be executed "at the same time."

```
Q      <= S nand NQ ;
NQ     <= R nand Q ;
```

These two statements will be called concurrent statements, and the compiler/simulator will guarantee that the order of their writing is not relevant. More precisely, figure 2.18 shows the way the compiler/simulator translates these statements into processes.

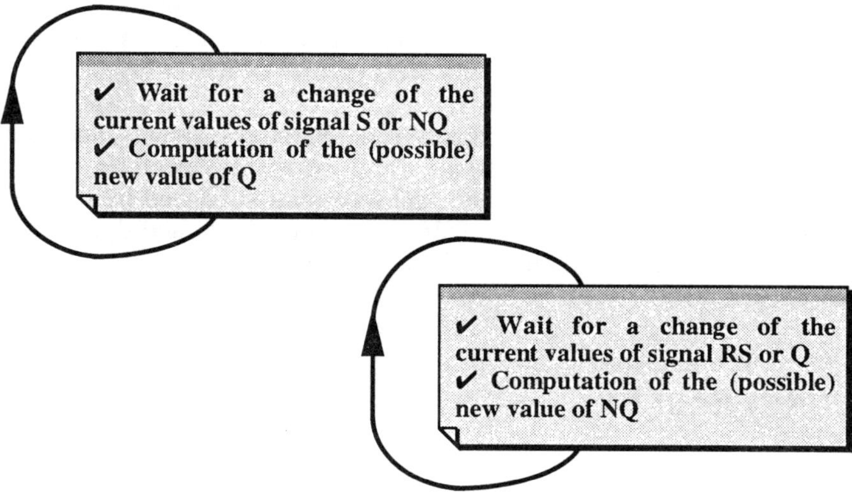


**Figure 2.18.**        **Simulator's View of the Latch**

These processes are executed independently of each other. This example will be discussed in more detail.

## 2.6.2. Signals and Ports

In boards or integrated circuits, it is easy to distinguish two kinds of wires: those interconnecting components and which are local to the architecture and those having access to the outside, the ports. A port has a name, which is known from the outside and a direction: **in, out** or bidirectional (**inout**).

There is no difference in the nature of ports or signals: signals transit through ports as shown in figure 2.19.

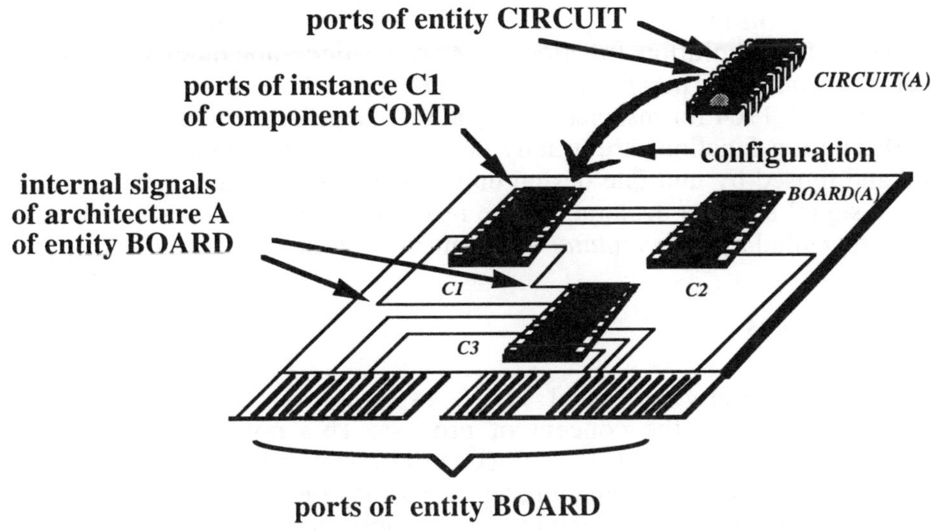


**Figure 2.19.**		**Signals and Ports**

### 2.6.3. Resolution Function

On a board as well as within integrated circuits, connecting two (or more) outputs together is delicate. What happens if they do not have the same value?

Depending on the target technology (TTL, ECL, etc.), the resulting values of conflicts are known and such connections can be made.

When several outputs are connected together in VHDL, the compiler requires the existence of a *resolution function*. Such a function is systematically called to solve conflicts. Its role is to compute at run-time the resulting value from the interconnected sources.

It is possible for the designer to write his or her own resolution function. Indeed, when using a conventional logic type (for example, the nine state logic values of package STD_LOGIC_1164), several associated resolution functions are also proposed (exported) by the package. They are usually defined together with an overloaded logic operator (and, or, etc.) working on the corresponding logic type.

Thus, VHDL offers a mechanism, the resolution function, which allows for the modeling of all technologies. Unlike proprietary simulators, the description of the behavior in the case of conflicts is explicit.

With guarded blocks, which are studied further in paragraph 2.6.6, resolution functions are another way of modeling bus disconnection mechanisms. The semantics of value 'Z', also called *three-state* or *high-*

*impedance*, is used for that purpose. This semantics is entirely deduced from the associated resolution function: a 'Z' value does not modify the resulting value of a conflict. Therefore, using a logic type where 'Z' is defined (package STD_LOGIC_1164 for instance), a signal may appear as disconnected.

Resolution functions are usually defined in packages and therefore may be used and reused by multiple design units. When present in a signal declaration or in a signal subtype declaration, the resolution function is automatically and implicitly called *by the simulator* each time a conflict has to be solved.

### 2.6.4. Process

All the descriptive power of VHDL, as far as concurrency (parallelism) is concerned relies on the concept of process. This notion has already been introduced in paragraph 2.6.1. Each concurrent statement is made up of one or more processes. During simulation, a process may only be in one of two states:
- either it is executing (running) as a subprogram
- or it is waiting for changes on signal values *(event)* and is blocked on a synchronization statement: **wait** statement.

When explicit, the syntax of a process is

```
[ process_label : ] process [ ( sensitivity_list) ] [ is ]
              process_declarative_part
          begin
              { all_sequential_statements }
          end process [ process_label ] ;
```

A process may be seen as a subprogram, which is only called once, at the beginning of the simulation, and loops. It runs to the next **wait** statement, waits for the corresponding event, and runs again to the next **wait** statement. When arriving at its final **end** reserved word, it instantly restarts after its **begin** reserved word.

The **wait** statement offers several different forms of synchronization:
- Waiting an event on a signal in a list (sensitivity list)
- Waiting until a condition becomes true
- Waiting for a certain period
- Waiting for any combination of the three previous forms

In the previously introduced example of latch, the two concurrent statements

```
Q     <= S nand NQ ;
NQ    <= R nand Q ;
```

have two equivalent processes:

```
process
begin
-- Sequential signal assignment
Q <= S nand NQ;
wait on S, NQ;
end process;

process
begin
-- Sequential signal assignment
NQ <= R nand Q;
wait on R, Q;
end process;
```

### 2.6.5. Other Concurrent Statements

Concurrent statements are used for describing behavior or structure of architectures. The order in which they are written does not modify the description. With the exception of component (and direct) instantiation,[7] each concurrent statement has an equivalent process.

Structural descriptions are based on the two kinds of instantiations, component instantiation and direct instantiation, which have respectively been described in paragraphs 2.2.2 and 2.2.5. They may be mixed, within the same architecture, with behavioral statements.

A concurrent procedure call statement, also previously seen in paragraph 2.2.7, is one of them. Under certain conditions, this statement may appear to be very close to the instantiation of a model made up of a single process. Speaking in terms of equivalent process, a concurrent procedure call statement may be translated into the process below containing a *sequential* call statement of the same procedure. Therefore the concurrent statement

```
CALL_TO_THE_PROCEDURE(S1, S2, X);
```

where S1 and S2 are parameters of class signal, is equivalent to the explicit process:

```
process
begin
        CALL_TO_THE_PROCEDURE(S1, S2, X);
        wait on S1, S2;
end process;
```

---

[7] The instantiation of a component described using $n$ processes is equivalent to $n$ processes!

The semantics of the process statement implies that the procedure is called each time an event occurs on one of its **in** or **inout** parameters of class signal.

Dataflow statements allow a behavioral description to be written as a set of concurrent signal assignments. Some forms of these statements are given below:

```
signal <= function(other signals, parameters);
signal <= signal <operator> signal;
```

Selected (a unique branch is selected depending on a value) or conditional forms (the execution only occurs when a condition is true) are also proposed; their syntax is given in appendix 8.1.

The equivalent process for the simple form of signal assignment follows:

```
process
begin
        signal <= expression_containing_signals,
        wait on signals_of_the_expression;
end process;
```

The embedded sequential signal assignment is therefore reexecuted each time an event occurs on signals appearing on the right-hand side of the concurrent assignment.

In the same spirit, the concurrent assertion can be easily defined by relying on the sequential assertion statement whose syntax is given in section 2.7. Its syntax is the same in both sequential and concurrent domains, and its equivalent process is given below:

```
process
begin
        assert condition_containing_signals report message severity severity_level;
        wait on signals_of_the_condition;
end process;
```

The main use of such a *passive* statement (no event on other signals may be created by such an assignment) is for dynamic constraint checking on signals (causality, set-up, hold, pulse width, etc.). There is therefore no synthesis semantics for it, and this statement is simply ignored by synthesis tools.

A final concurrent statement is the **generate** statement which "macro generates" concurrent statements. It is the means of describing potential structures (using the conditional form, see 3.5.6.2) or highly regular structures (using the iterative form, see 3.5.6.1). A synthesis tool, respectively a simulator, must elaborate **generate** statements before the synthesis, respectively simulation, is performed. A **generate** statement does not add any functionality to the behavior of a model, but it provides much more generality and flexibility to the description.

## 2.6.6. Guarded Blocks

Guarded blocks are a way of putting signal assignments together and of executing them only when a condition, the *guard condition*, is true. Indeed, two main applications of guarded blocks arise:
- For synchronous descriptions, the set of synchronous statements are placed within a guarded block. The guard condition is related to the clock (rising, falling, level, etc.).
- Choosing the right guard condition, bus disconnection mechanisms may be described. It is important to note that guarded blocks are not the only way to model bus mechanisms, a convenient logic type (with value 'Z' as in package STD_LOGIC_1164 presented in section 6.2) may also be used.

## 2.7. SEQUENTIAL DOMAIN

It is possible to imagine a Hardware Description Language only offering structural description (instantiation) and dataflow statements. UDL/I, the Japanese standard is a perfect example of this. Nevertheless, certain parts of a system are usually easier to describe using sequential statements: these statements are executed in sequence. This is due to the very nature of the description (do that, wait for that, then do this) or more simply to the programming habits of the designer (C, Ada, etc.).

In VHDL, sequential descriptions are allowed in processes and subprograms. These statements are for the main part inspired from classical programming languages (**if, case, loop, while**, etc.). The syntax of these structured sequential statements (they contain other sequential statements) is given in appendix 8.1.

The signal assignment symbol (<=) differs from the variable assignment (:=). Indeed, as previously seen in section 2.4, their semantics is quite different. Signal assignment syntax is:

```
[ label : ] target <= [ transport | inertial ] waveform ;                  -- VHDL'87
[ label : ] target <= [ transport | [ reject pulse_width] inertial ] waveform ;   -- VHDL'92
```

The **after** clause of waveforms, waveforms with multiple clauses as well as the **reject** clause are not taken into account for synthesis purposes. Since this statement is the most basic one of VHDL, it illustrates very well the following remark. VHDL was not designed as a language for synthesis, and a synthesis semantics has only afterward been associated with the language, when possible.

Two specific statements, **assert** and **report** (*VHDL'92*), are mainly used for displaying messages during simulation. They are ignored by synthesis tools.

```
assert condition_containing_signals [ report message_to_display ] [ severity severity_level ] ;
report message_to_display [ severity severity_level ] ;
```

The synchronization statement, the **wait** statement, has already been presented in paragraph 2.6.4. It waits until an event occurs on a list of signals, a condition becomes true, or a time-out occurs, and it has the following syntax:

```
wait on [list_of_signals] until [condition] for [a_given_time_period] ;
```

The update of signal drivers is performed when encountering this statement. The (event-driven) simulator advances the simulation time to the next event, when all processes are stopped on such a statement: the whole simulation mechanism relies on this statement.


## 2.8. ATTACHED CHARACTERISTICS


Adding (attaching) information to an object may be useful. This information is usually devoted to specific back-end tools of the design methodology: synthesis tools are among them. There are two different ways of introducing this information in the VHDL code, syntactic comments, or attributes:
- Syntactic comments consist of defining a syntax and associating semantics to a comment (which in VHDL begins with "--" and is closed at the end of the line):

```
-- Synthesis on
```

    may mean that the synthesis process has to be started here. This notion of syntactic comment has to be known since it is necessary for the use of certain synthesis tools. Conceptually, this practice is frustrating: it consists in defining a proprietary language (with its own syntax) within VHDL. Therefore, its use should disappear in the next few years.

- Attributes are an intrinsic notion of the language. An attribute is a characteristic, which may be attached to an object or to a group of objects. This notion of object is very wide and includes statement labels as well as signals, entities or subprograms. The complete list of attributable objects is given in section 5.1 of the LRM.

If some of them are predefined (they are part of the environment), the designer may define his or her own ones. Such user-defined attributes may only be values: for example, attribute TIMEOUT of type TIME may be defined on signal S with a value of 10 ns. Attributes are a smarter way than syntactic comments to enter into the language the information needed by back-end tools. The information for controlling the synthesis process may potentially be expressed using attributes of a predefined package.

## 2.9. PREDEFINED ENVIRONMENT

One of the main criticisms of VHDL in its early stages was its too high degree of generality and the weakness of its predefined environment. Through the years, things have evolved by the addition of new features (*VHDL'92*), by the standardization of specific packages (STD_LOGIC_1164, math package, etc.), and by the definition of standard modeling practices (VITAL).

### 2.9.1. Predefined Environment of the Language Itself

This environment is defined within the language reference manual (LRM) and therefore is part of the language: any IEEE standard compliant implementation has to propose it. It consists of
- A set of useful predefined types within the package STANDARD. Types TIME and BOOLEAN are among them. No **use** clause is necessary to refer this package: it is implicitly referenced.
- I/O primitives for types with ASCII representation (string of characters, BIT, integers, reals, etc.) exported by package TEXTIO. This package has no interest in a synthesis process (and its primitives are not synthesizable) but is useful in the simulation environment for allowing pattern generation and traces.
- Several families of operators predefined in VHDL: logic (and, or, etc.), relational (=, <, etc.), addition (+, -, etc.), sign (+, -), multiplication (*, /, etc.), and miscellaneous (**, ABS, etc.) operators to which have to be added the shift and rotate operators of *VHDL'92* (sll, rol, etc.).
- Some attributes predefined in VHDL. They are part of the predefined environment and are very useful to the designer, both for simulation as well as for synthesis purposes. They may be classified according to their kind (value or signal) and the objects to which they apply (arrays, signals, types, etc.). Their complete presentation is outside the scope of

this book, but it would seem useful to take an example of attributes of each kind to highlight their use.

Among predefined attributes on arrays, 'LENGTH is very valuable: A'LENGTH, which is an expression, returns *a value*, which is the size of the array A. Used within a subprogram, this attribute allows a great flexibility of the parameters and therefore the design of a more general, and thus reusable, algorithm.

Attributes on signals are beyond all question the most frequently used ones in VHDL descriptions. S'LAST_EVENT, for example, returns the period of time since the last event occurred on signal S. This returned *value* is therefore useful for checking timing constraints such as set-up. Belonging to the other kind of attributes, the attribute S'DELAYED(10 ns) returns a *signal* (and not only a value) which is the signal S, delayed of 10 ns. This opens the door to powerful timing checks.

Some attributes (such as 'EVENT) may also be used by synthesis tools for recognition of clocks or resets. This is only a ("proprietary") synthesis interpretation of them, and attributes on signal have no semantics concerning this use.


## 2.9.2. Standard  Environment

This has to be distinguished from the predefined environment. It is not part of the language itself but nevertheless its definition has followed a similar standardization process with a final ballot. It is widely accepted, and any VHDL vendor has to propose it.

It appears as a set of standardized packages that can be referenced with a use clause:

- Package STD_LOGIC_1164, also called "nine-state package," defines a logic type and associated operators and resolution functions. Well accepted by tool and model library vendors, it is hard-wired, and therefore optimized in many tools. Furthermore, it is one of the basic elements of VITAL (VHDL Initiative Towards ASIC Libraries) and therefore contributes to a good compatibility of descriptions using it with model libraries. The synthesis semantics of the nine states has been (partly) standardized and will be discussed in section 6.2.
- Arithmetic packages, based on type BIT or on type STD_ULOGIC, are also part of the standard environment. They offer many arithmetic operators, functions, and procedures (with or without signs) that are directly synthesizable. They will be presented in paragraph 6.3.2.
- The problem of the expression of synthesis constraints has been discussed for a long time within IEEE working groups. No real standard exists.

### 2.9.3. "Proprietary" Environment

The result of a synthesis process is conceptually a netlist of components. These components are predefined within a "proprietary" model library. There is no standard here, and the contents of such libraries as well as their abstract level are rather different. Some vendors even propose different libraries, such as the following:

- libraries accurately characterized and integrating a back-annotation mechanism, they allow an expensive but accurate simulation
- libraries more roughly described with timing values (elementary time unit), they nevertheless allow a quick check of the logical behavior of the model
- libraries that are functional and that target potentially acceleratable primitives

Other options may be offered and are part of the usual environment available on VHDL synthesis platforms.

# 3. MAPPING VHDL TO HARDWARE

*This chapter may be considered to be a software approach of synthesis. Designers more interested in a hardware approach may read chapter 4 before this one.*

## 3.1. SYNTHESIS MODELING STYLE

The synthesis modeling style is mainly characterized by reducing the VHDL possibilities to a subset in which:
- delay expressions (**after** clauses, **wait for** statements) are ignored
- certain restrictions on the writing of process statement occur
- only a few types are allowed
- description is oriented towards synchronous styles by expliciting clocks

This subset and its synthesis semantics is detailed in the following pages.

## 3.2. VHDL TYPES

As previously seen, VHDL objects such as constants, signals, and variables are declared with a type or a subtype. A type or a subtype defines the set of authorized values and the operations it is possible to perform on these objects. For example the predefined type BIT is declared by an enumerated type listing two possible values ('0' and '1'). Functions associated with this type are **not, and, xor**, etc.

## 3.2.1. Supported  Type  Declarations

Modeling for synthesis only supports a few types: enumerated type, integer, one-dimensional array, and record types.

### 3.2.1.1. Enumerated  types

Enumerated types are defined by giving their complete list of values. Each element of this list can be either an identifier different from a reserved word (such as: FIRST_ITEM ) or a character literal (such as '1' ).

```
type STD_ULOGIC is ('U', 'X', '0', '1', 'Z', 'W', 'L', 'H', '-');
signal S : STD_ULOGIC;

type STATE_TYPE is (HALT, READY, RUN, ERROR);
variable STATE : STATE_TYPE;

type CODE_TYPE is (NUL, '0', '1');
function CODE (C : in INTEGER) return CODE_TYPE;
```

**Remarks:**
- In the example above, the definitions of both types STD_ULOGIC and CODE_TYPE include enumeration literals '1' and '0' which are said to be overloaded. VHDL authorizes such duplications and the compiler is able to recognize the type of the literal. However, to solve ambiguous statements, the use of qualified expressions is required, such as: TYPE_NAME'(EXPRESSION). This qualification should be widely recognized by synthesis tools and should not imply any extra hardware: this information is only used at the compilation level to help the compiler.
- Types STD_ULOGIC and STATE_TYPE are both enumerated types; nevertheless the level of abstraction of STATE_TYPE is higher than that of STD_ULOGIC and therefore STATE_TYPE need to be encoded.

The main synthesis issue with enumerated type is the encoding: there is no consensus for encoding enumerated types. Very often, the enumeration values are encoded by default into bit vectors whose length is the minimum number of bits required to code the number of enumerated values. This policy is applied to the predefined enumerated type BOOLEAN, where FALSE and TRUE are respectively encoded as '0' and '1'. This solution is not systematically chosen, especially when the enumerated type is used for defining the state of a finite state machine. In that case, another way of encoding has to be chosen to optimize the result.

The use of enumerated type in modeling is strongly recommended and is one way of remaining at an abstract level and relying on the synthesis tool to choose the right strategy when optimizing and encoding.

### 3.2.1.2. Integer   types

Integer types are defined as subranges of a more general anonymous built-in type and some bounds have to be declared.

```
type LENGTH is range 0 to 1000;
type BYTE_INT is range -128 to 127;
type MY_INTEGER is range -2147483647 to 2147483647;
```

The number of bits required to represent integer values will be computed from this range. In many cases, if the integer is negative, it is assumed to be encoded as a two's complement bit vector.

**Remarks:**
* Bits of an integer type value are not directly accessible. No assumption is made on the location of these bits. Specific functions translating an integer into a bit vector are defined and exported by packages provided in the synthesis tool environment. Standardized packages are presented in chapter 6 devoted to the synthesis environment.
* Integer types defined by the user are incompatible but "closely related" to the type INTEGER of package STANDARD. The only way to ensure compatibility is to use a type conversion to change the type of expression. This possibility does not imply extra hardware, it is only a workaround for dealing with the strong typing of VHDL. Another recommended practice is to define subtypes of standard INTEGER, so that the objects declared with these subtypes remain compatible.

```
-- Within an architecture declarative part
type MY_INTEGER is INTEGER range 0 to 10_000;
signal MY_INT : MY_INTEGER;
subtype SMALL_INTEGER is INTEGER range -15 to 16;
signal S_INT : SMALL_INTEGER;

...
-- Within a process declarative part
variable INT : INTEGER:= 2;

...
S_INT <= INT;
INT := S_INT;

...
MY_INT <= MY_INTEGER(INT);
INT := INTEGER(MY_INT);
```

### 3.2.1.3. Multi-dimensional array types

Multi-dimensional array types are supported by VHDL for defining an indexed set. Any previously defined type may be an element of these arrays (except unconstrained arrays). Nevertheless only one-dimensional arrays are usually allowed by synthesis tools. This arbitrary restriction should be removed in the future. One way of getting around this unsupported feature is to declare two one-dimensional array types instead of one two-dimensional array:

```
type WORD is array (31 downto 0) of BIT;
type RAM is array (1023 downto 0) of WORD;
```

Instead of

```
type OTHER_RAM is array (1023 downto 0, 31 downto 0) of BIT;
```

Objects of types RAM and OTHER_RAM are the same size. The way of addressing an elementary BIT makes the distinction between them. In the first case R1(x)(y), and in the second case R2(x, y) index the same element. The reason multi-dimensional arrays are not supported by synthesis tools is the difficulty of deciding the hardware to compute the item address. In the general case, multipliers and adders have to be assumed, and this is a very expensive layout.

The ability to declare constrained or unconstrained array types is given by VHDL. Unconstrained array types are reserved for defining generic array types from which constrained subtypes or objects are derived:

```
architecture A of E is
        subtype INT is INTEGER range 0 to 7;
         -- Unconstrained array type
        type INT_VECTOR is array (NATURAL range <>) of INT;
        -- Constrained array subtype
        subtype WORD is INT_VECTOR(7 downto 0);
        signal S1, S2 : WORD;
begin
        P: process
            variable V : INT_VECTOR(1 to 2);
        begin
            S1(7 downto 6) <= V;

            ...
```

Another important advantage of using such unconstrained arrays is that the index range of each parameter declaration (generic parameter, port parameter, subprogram parameter) is provided together with the object itself and can be determined via attributes. This provides the ability to write more general subprograms or components, which are independent of any bounds.

```vhdl
entity FIRST_ONE is
        port (V : in BIT_VECTOR;
            RESULT : out BIT_VECTOR);
begin
        assert (V'length = RESULT'length)
            report "V''length and RESULT'length are not consistent"
            severity ERROR;
end FIRST_ONE;

architecture A of FIRST_ONE is
begin
        P : process (V)
            alias LV : BIT_VECTOR (V'length downto 1) is V;
            variable R : BIT_VECTOR(LV'range);
        begin
            for I in LV(LV'range) loop
                if LV(I)='1' then
                    R(I):= '1';
                    for J in I-1 downto 1 loop
                        R(J) := '0';
                    end loop;
                    exit;
                else R(I):= '0';
                end if;
            end loop;
            RESULT <= R;
        end process;
end A;
```

The previous example needs some comments:
- An **alias** statement is useful here to force a descending range to vector V, because the declaration of the actual parameter is in this case unknown. This declaration does not imply specific hardware.
- All the objects, constants, variables, and signals must be declared with constrained subtypes (to be able to compute their size during compilation). Here, the local variable R is constrained by the range inherited from the alias declaration LV whose range is itself inherited from the function parameter V. Such properties are applied to the simulation and to the synthesis domain. Moreover, in synthesis applications, the object sizes are computed in terms of hardware.
- An **assert** statement is included in the concurrent statement part of the entity. There is no hardware counterpart of this statement; however, this is a useful checking statement during the elaboration run time.

### 3.2.1.4. Record  types

Record types are useful for defining a set of various types (any previously defined one, except unconstrained array types). A record element is addressed by its name, which can be translated by the synthesis tool as a constant offset value or flattened in some cases (the notion of record no longer exists, the item address is computed only once and becomes hardware coded).

```
architecture A of E is
      type CODE_TYPE is (NONE, DATA, STATMT);
      type ITEM_TYPE is record
          CODE : CODE_TYPE;
          INT : INTEGER;
      end record;
      signal S1, S2 : ITEM_TYPE;
begin
      process
          variable V : ITEM_TYPE;
      begin
          S1 <= V;
          V:= S2;

          ...

          S2.INT <= 0;
          V.CODE:= S1.CODE;

          ...
```

## 3.2.2. Predefined  Types

STANDARD and TEXTIO are two packages described in the LRM. TEXTIO is dedicated to ASCII file inputs/outputs, and this package is not needed for synthesis. Package STANDARD defines types such as INTEGER, BIT, BIT_VECTOR, BOOLEAN, CHARACTER, and subtypes NATURAL, POSITIVE, which are synthesizable. Only **file**, **access**, and type REAL are not synthesizable: the former have no hardware counterpart, and for the latter no agreement for encoding a real value exists (the number of bits are necessary to encode exponent and mantissa).

## 3.2.3. IEEE  Strongly  Recommended  Types

A package defining multiple logic type values was successfully balloted in 1992: STD_LOGIC_1164. It standardizes a nine-state logic value with a simulation semantics, and its semantics is currently extended to cover synthesis needs. This package is supported by all synthesis tools and presented in section 6.2. Only a quick overview of the synthesis interpretation of the enumerated type is described here.

This package defines the following logic type:

```
type STD_ULOGIC is (    'U',    --    Uninitialized
                        'X',    --    Forcing Unknown
                        '0',    --    Forcing 0
                        '1',    --    Forcing 1
                        'Z',    --    High Impedance
                        'W',    --    Weak Unknown
                        'L',    --    Weak 0
                        'H',    --    Weak 1
                        '-'     --    Don't care   );
```

'X', '1', and '0' are said to be strong values and dominate the weak values 'W', 'L', 'H' which themselves dominate 'Z'.

'1' and '0' can be interpreted as tied to power and respectively tied to ground.

Values 'U', 'X', and 'W' are said to be metalogical values. They have mainly a simulation function and no obvious hardware meaning. The value 'U' is the leftmost literal in the enumerated STD_ULOGIC definition, and is the default value for the initialized variables or signals prior to being explicitly assigned. The values 'X' and 'W' represent the states that cannot be determined by the simulator. 'W' having a lesser effect can be forced to "0', '1', and 'X'.

Since the synthesis tools are not yet able to differentiate between strengths, values 'L' and 'H' have no standard synthesis semantics.

'Z' may be used in simulation as a result of a resolution function when no drivers are activated. For synthesis, specific hardware is implied: when the conditional assignment of scalar value 'Z' is used, the target of the assignment becomes the output of a tristate buffer. In the following examples, the signals SOURCE and CONDITION respectively represent the input and command of the tristate buffer. The two statements below are equivalent, TARGET1 and TARGET2 being identical.

```
-- example 1
TARGET1 <= SOURCE when CONDITION else 'Z';

-- example 2
with CONDITION select TARGET2 <= SOURCE when TRUE, 'Z' when FALSE;
```

This writing is straightforward due to the fact that signal CONDITION is assumed to be of type BOOLEAN. If the type of this signal was STD_ULOGIC, a signal assignment dealing with '0' and '1' values (instead of TRUE and FALSE) may be written as follows:

```
-- example 1
TARGET1  <=   SOURCE when CONDITION ='1' else
              'Z' when CONDITION ='0' else
              'X';
```

```
-- example 2
with CONDITION select TARGET2 <= SOURCE when '1',
                                 'Z' when '0',
                                 'X' when others;
```

In this last example, what is the meaning of 'X', and what hardware is implied? At the simulation step, if 'X' is raised, that means an error, but in the synthesis domain the 'X' value is interpreted as "Don't care" value.

'-' is said to be the "don't care" value, and has been introduced to obtain a better optimization result of the hardware provided by the synthesis process. In general, it is interpreted by the synthesis tool as either '0' or '1'.

In the following example, the value of signal A has to be assigned to signal X when SEL equal "001" and the value of signal B has to be assigned to signal X when SEL equal "010". Other values of SEL are not relevant, i.e., the designer does not care of them.

Using a **case** statement, architecture A simply forget the other values of SEL. Since the **others** clause is mandatory, keyword **null** may be used:

```
architecture A of DONT_CARE is
        signal SEL : BIT_VECTOR(1 to 3);
        signal A,B, X : BIT;
        ...
begin
        process (A, B, X)
        begin
            case SEL is
                when "001"   =>  X <= A;
                when "010"   =>  X <= B;
                when others  =>  null;
            end case;
        end process;
end A;
```

This description implies memorization elements since the hardware counterpart is sequential: signal X is not assigned whatever is the **case** branch. This is shown in figure 3.1.

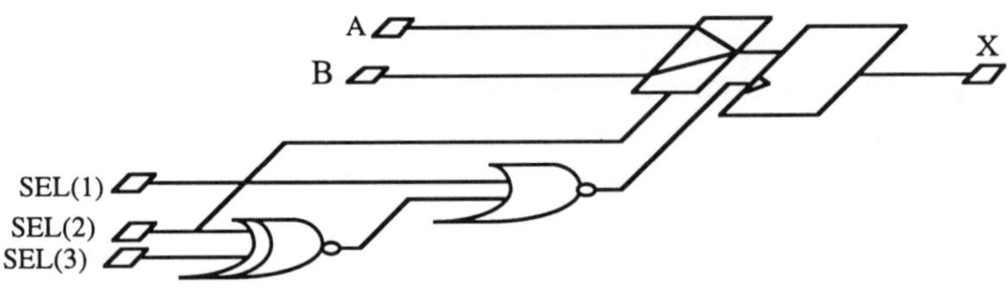


**Figure  3.1. Sequential Hardware Inferred by Architecture A**

An obvious solution is to force the value of signal X in the **others** clause. Architecture B uses the value '0' for this purpose (value '1' was possible, too).

```
architecture B of DONT_CARE is
        signal SEL : BIT_VECTOR(1 to 3);
        signal A,B, X : BIT;
...
begin
        process (A, B, X)
        begin
            case SEL is
                when "001" => X <= A;
                when "010" => X <= B;
                when others => X <= '0';
            end case;
        end process;
end B;
```

The result is better in terms of inferred hardware (see figure 3.2), but forcing signal X to a given value has a cost.

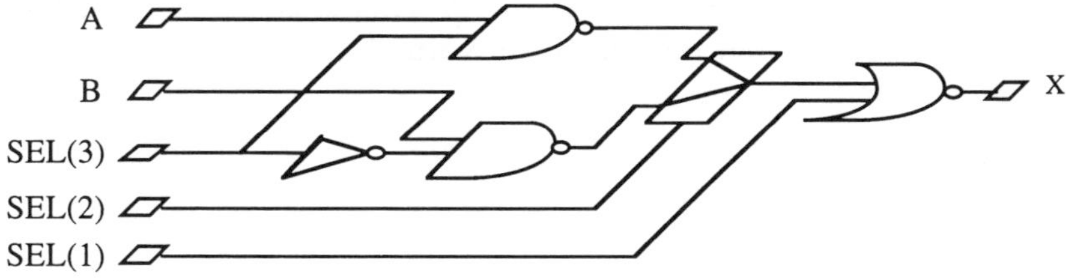


**Figure 3.2. Architecture B, Value '0' Is Used as "Don't Care" Value**

The optimal solution may be finally reached by letting the synthesis tool free to perform optimizations. Architecture C, using the "don't care" value ('-') and thus type STD_ULOGIC instead of type BIT, reaches this goal. The corresponding hardware counterpart is presented in figure 3.3.

```
library IEEE;
use IEEE.STD_LOGIC_1164.all;

architecture C of DONT_CARE is

        signal SEL : STD_ULOGIC_VECTOR(1 to 3);
        signal A,B, X : STD_ULOGIC;
...
begin
```

```
process (A, B, X)
begin
    case SEL is
        when "001" => X <= A;
        when "010" => X <= B;
        when others => X <= '-';
    end case;
end process;

end C;
```



**Figure 3.3. Architecture C Using a "Don't Care" Character**

When a "don't care" value is involved in a comparison, simulation and synthesis semantics differ. For a simulator, a "don't care" is a value systematically different from value '0' or '1'. This is completely different from the synthesis semantics that interprets the value '-' as a wild card. This issue is discussed in section 6.2.

### 3.2.4. Unsupported Types

Some datatypes are not useful for synthesis purposes (for example, all physical types defined by the designer are not supported). Predefined physical type TIME is not supported. Even timing constraints for synthesis are not expressed in VHDL with time expressions. They are external constraints given to the synthesis tool using attributes or proprietary mechanisms. Therefore, **after** and **reject** clauses are not interpreted and no synthesis tool can guarantee that a signal will change after an exact value of time.

If these clauses are used in the VHDL description, synthesis tools ignore them and the hardware result may be inconsistent with the input description. This implies that a correct modeling style for synthesis does not use any time expression.

This encompasses the **wait for** time-expression statement. In simulation the process is suspended for the time value defined by the **for** expression. During this period, all signals keep their new values. In the synthesis domain, it is impossible to infer hardware with such results without providing hardware that is too complex and overspecified (such as timers).

### 3.2.5. Subtypes

Subtypes are useful for synthesis. These are ways of accurately describing the range of variables or signals. At the simulation step, subtypes provide powerful possibilities for using type checking. Subtypes inherit all the operators defined from their base types. For example a function operating on type BIT_VECTOR can be used with any subtype of BIT_VECTOR as long as the values remain within the right range.

In the synthesis domain, the range defines the number of bits for encoding, and it is therefore strongly recommended to frequently use subtypes to provide optimized hardware:

```
constant N : NATURAL := 7;
subtype COUNT is NATURAL range 0 to N;
subtype COUNT2 is NATURAL range 0 to 2*N;
signal WORD : COUNT;
signal WORD2 : COUNT2;
```

Such kinds of subtypes may be used for creating a counter modulo 8 encoded as three bits for signal WORD and four bits for signal WORD2. Then the following encoding of the next statement is optimal;

```
WORD <= WORD2 / 2;
```

## 3.3. VHDL  OBJECTS

This section describes the VHDL objects and their hardware counterpart. This section only contains information on constants, variables, and signals. Files, which are objects out of the synthesizable domain, are not studied in the framework of this book.

### 3.3.1. Constants

A constant value is computed only once. The synthesis process accepts constants of any kind of synthesizable types. Does a constant declaration infer hardware?

Truth tables, ROMs can be modelled using constants. In that case, such declarations do not generate any hardware:

```
type TAB2 is array (BIT, BIT) of BIT;
constant AND_TAB : TAB2 := (('0','0'), ('0', '1'));
```

On the other hand, if this constant is used in a signal assignment, as in the following example, hardware is inferred.

```
signal Z, A, B : BIT;
...
Z <= AND_TAB(A, B);
```

Indeed constant *declarations* do not generate hardware. Therefore, it is the use conditions of constants that determine the inferred hardware. When or where does a constant have to be used in a model? Coming back to its definition: a constant is an expression, so it may be found in the same place as expressions.

- On the right-hand side of a signal assignment:

```
constant COD1 : BIT_VECTOR := X"EA";
constant MASK : BIT_VECTOR := "01111111";
signal V, R, A: BIT_VECTOR(7 downto 0);
...
begin
...
        V <= COD1; R <= A and MASK;
```

- In an **if** or **case** statement expression (sequential domain, i.e., the order in which these statements are written is important):

```
if S = CST1 then ...

case VALUE is
        when CST_N => ...
```

- In a conditional concurrent statement:

```
Z <= CST1 when A='1' else S when B= CST2 else CST3;
```

Synthesis tools use constant coding for minimizing the inferred hardware. A nonnegligible part of a design can be described using constants. The design may perform complex computations during the initialization phase and the elaboration computes and propagates constant values. Moreover the synthesis step optimizes the resulting hardware by propagating the constant values and simplifying it when possible.

```
constant ROM : ROM_TYPE := READ("rom_file.dat");
signal WORD5 : BIT_VECTOR (3 downto 0);
begin
        WORD5 <= ROM (5);
```

In the previous example, the WORD5 signal becomes constant due to the fact that ROM and its index value are constant. The synthesis tool causes the signal WORD5 to disappear by coding propagation.

VHDL also defines deferred constants, which are declared in the specification package and whose values are defined in the body of the same package. These constants are not supported by all synthesis tools. There is no major difficulty in implementing them if the elaboration phase is well executed.

At the end of this chapter, the notion of a generic parameter is studied as a generalization of the constant usage.

The notion of constant can be hidden. For example the **open** value for the **in** or **inout** port of a component can be considered as a **constant** signal value inside the architecture of a model. In this case the default value must be given in the entity declaration. This is also the schema for the **open** value for the **in** and **inout** port of component instantiation, and in this case the default value must be defined in the component declaration. The notion is the same for all default values defined in subprogram headers and when used during subprogram calls, can be seen as constants.

### 3.3.2. Variables Versus Signals

As described in the previous chapter, signals and variables have different behavior during simulation. Therefore, the hardware resulting from synthesis of variables or signals is different: either nothing, or wire, or memory elements.

- Example 1 using variables:

```
entity MEMO_ONE is
      port (DATA        : in BIT_VECTOR (1 downto 0);
            CLOCK       : in BIT;
            Z           : out BIT);
      constant K1 : BIT_VECTOR := "01";
      constant K2 : BIT_VECTOR := "10";
end MEMO_ONE;

architecture A of MEMO_ONE is
begin
      process (CLOCK)
            variable A1,A2 : BIT_VECTOR(DATA'range);
            variable A3 : BIT;
      begin
            if CLOCK = '1' and CLOCK'event then -- "and CLOCK'event" is not required for
            -- simulation. It is introduced to be compatible with certain synthesis tool restrictions.
                  A1 := DATA and K1;
                  A2 := DATA and K2;
```

```
                A3 := A1(0) or A2(1);
                Z <= A3;
            end if;
        end process;
    end A;
```

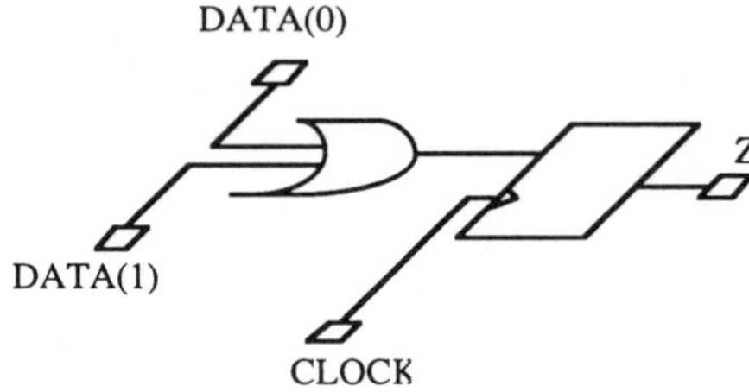



**Figure 3.4. Synthesized Result for Architecture A of MEMO_ONE**

In the previous example, the range of the port DATA is the range of a bit vector constrained to a two-bit length. The type BIT_VECTOR being defined as unconstrained, it is necessary for the synthesis to specify the real size of this port. By default, it is generally assumed to be a 32-bit length. This kind of range precision defines an implicit subtype.

The variables used in this example do not infer any hardware. Indeed, a conceptually equivalent process may be given without any variables, where all variables are replaced by their equations:

```
begin
    if CLOCK = '1' and CLOCK'event then
    Z <= (DATA and K2) (1) or (DATA and K1)(0); -- WARNING : this is not VHDL code
    end if;
```

- Example 2 using signals:

```
architecture B of MEMO_ONE is
    signal A1, A2 : BIT_VECTOR(DATA'range);
    signal A3 : BIT;
begin
    A1 <= DATA and K1;
    A2 <= DATA and K2;
    process (CLOCK)
    begin
        if CLOCK = '1' and CLOCK'event then
            A3 <= A1(0) or A2(1);
            Z <= A3;        -- Value of A3 computed during previous clock cycle
        end if;
    end process;
end B;
```

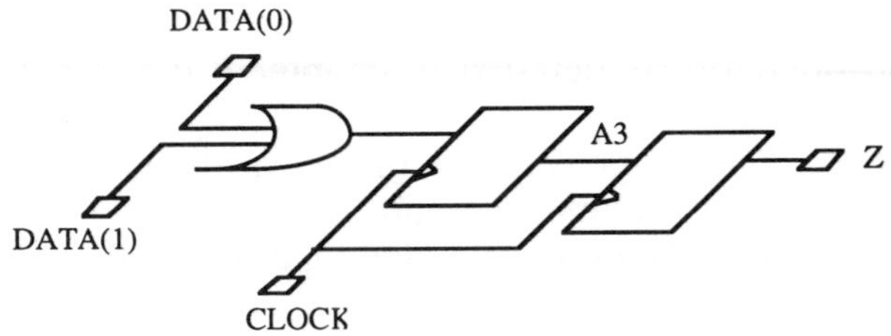


**Figure 3.5. Synthesized Result for Architecture B of MEMO_ONE**

As these two architectures are described differently, the synthesis results are different. Nevertheless, variables or signals A1 and A2 are processed in the same way as signals; they generate no hardware. However object A3, which is a variable in architecture A and a signal in architecture B, only generates a flip-flop in the second architecture. The A3 variable is taken as an intermediate program construction, but on the other hand the A3 signal creates a hardware intermediate storage.

### 3.3.3. Initial Values

In VHDL, there are three kinds of initial values: the default value inherited from type or subtype definitions, the explicit initialization which is given when the object is declared, and a value assigned using a statement at the beginning of a process.

The first and second cases are ignored by synthesis tools, which can generate inconsistencies between the behavior of simulation and synthesis results. It is therefore recommended to explicitly and systematically initialize variables and signals with specific statements. Very often this is achieved within part of the synthesizable code devoted to the set/reset treatment, and initialization must be found in this part.

If **out** ports, or **out** parameters of subprograms, have default values, they have the same behavior as an initial value for a signal or a variable. They are ignored by synthesis tools.

### 3.3.4. Arithmetic Operators

#### 3.3.4.1. Preliminary remarks

Some arithmetic operators are defined in the LRM and therefore these operators are overloaded for the multivalued logic standard package

STD_LOGIC_1164 and their associated arithmetic packages. The semantics of these overloaded operators is detailed in chapter 6 devoted to the synthesis environment.

Boolean and bit operators are part of the arithmetic operators. VHDL defines seven classes of operators. They are shown in figure 3.6 by increasing priority. As VHDL authorizes overloaded operators, the latter keep their priorities.

| Type | Operators | | | | | | Priority |
|---|---|---|---|---|---|---|---|
| Logical | or | and | nor | nand | xor | xnor | lowest |
| Relational | = | /= | > | < | >= | <= | |
| Shift | sll | srl | sla | sra | rol | ror | |
| Adding | + | - | & | | | | |
| Unary | + | - | | | | | |
| Multiplying | * | / | mod | rem | | | |
| Miscellaneous | ** | abs | not | | | | highest |

**Figure 3.6. Predefined VHDL Operators**

### 3.3.4.2. Logical operators

Logical operators, and **not** operator, accept operands of types BIT, BOOLEAN, and vectors with identical size for binary operators. The inferred hardware is a set of logical gates resulting from simplification of equivalent boolean equations. A convention widely accepted makes the Boolean value TRUE equivalent to the BIT value '1'.

```
signal S, X, Y : BIT_VECTOR(1 downto 0);
signal R, A, B, C : BIT;
signal T, D ,E ,F ,G : BOOLEAN;
...
begin
    S <= X and Y;
    R <= (A and B) and C;
    T <= D xor E xor F xor G;
```

**Figure 3.7. Logical Operations on Signals and Their Schematic Results**

### 3.3.4.3. Relational  operators

Relational operators always return a Boolean value, which is coded '0' or '1' and which is the result of the comparison of two operands of the same base type. For simulation, equality and inequality operators are implicitly defined for all types. The result is TRUE if the two operands represent the same value.

Relational operators are defined for all scalar types and one dimensional arrays of scalar types. The order of a scalar type is defined by its declaration, the type'LEFT being lower than the type'RIGHT.

For one-dimensional arrays, their relational order is defined by the lexical order, for example:

```
"0" < "1"            -- is true

"A" < "BC"           -- is true

"10" < "101"         -- is also true
```

**Remark:**
This relational operator cannot be used for comparing bit vectors coding numerical values. This operation has to be done with the arithmetic package detailed in chapter 6. Thus the result of the last statement "10" < "101" will be interpreted differently if unsigned or signed is adopted, and if the sign bit is the left bit or not.

### 3.3.4.4. Adding  operators

Plus and minus operators are defined for integer operands. All synthesis tools implement these operators and often have several capabilities that are used with respect to the constraints.

**Remark:**
Sometimes parentheses are a natural way of grouping a set of gates. In other words, two different expressions behave identically but infer different hardware. A way of clarifying the written code for simulation becomes a way of structuring the provided hardware. However, as the priority to solve area or timing constraints is given by the synthesis tool, this decision may overload the structuring required by parentheses in the code. For example, these three signal assignments infer different hardware even if the logical behavior is the same.

```
R <= A + B + C + 1;

S <= (A + B) + (C + 1);

T <= (((A + B) + C) + 1);
```

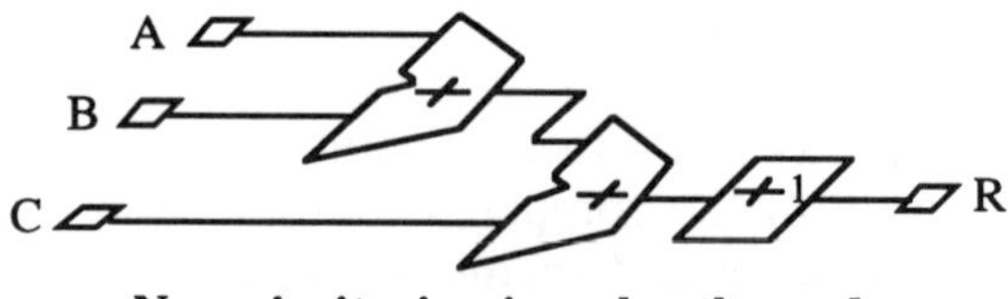

**No priority is given by the code**

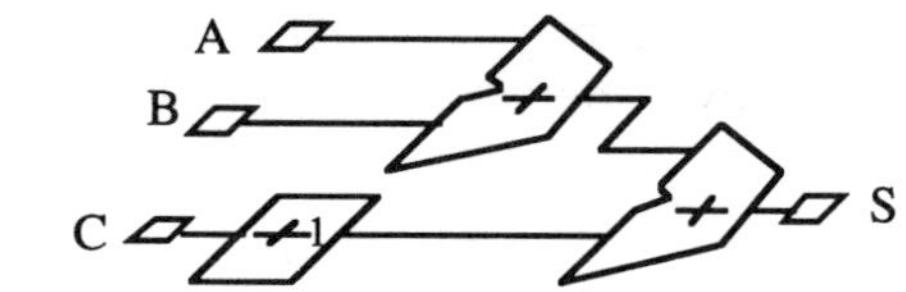

**A + B and C + 1 are simultaneous computed.**

**The priority to compute is given to A + B before C.**

**Figure  3.8. Parentheses and their Impact on the Inferred Hardware**

The concatenation operator is included in adding operators. It is predefined for all one-dimensional array operands. The result of this binary operator is always a one-dimensional array. Operands can be either an array or an element of this array.

```
architecture ...

        signal A, C  : BIT_VECTOR (2 downto 0);
        signal B,S,R : BIT_VECTOR (0 to 5);
        signal D : BIT;

begin

        A <= D & not D & D ;
        S <= A & B(0 to 2);
        R <= C (1 downto 0) & "000" & D;

        ...
```

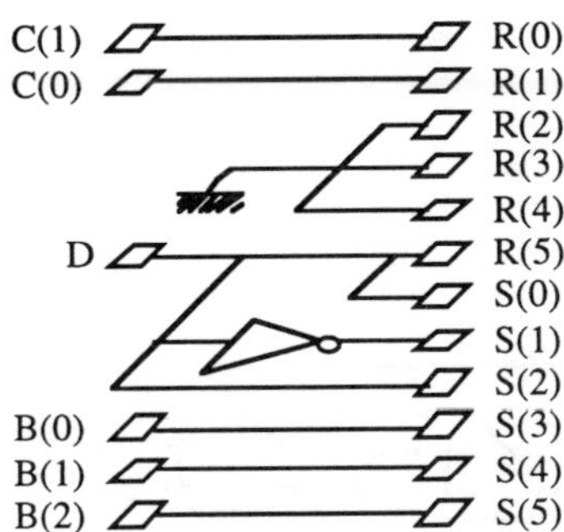

**Figure 3.9. Synthesized Result of the Operator "&"**

### 3.3.4.5. Shift operators

Shift operators are introduced into *VHDL'92*, are supported for BIT_VECTOR and are overloaded for STD_ULOGIC_VECTOR in the STD_LOGIC_1164 package. The shift and rotate operators are **sll, srl, sla, sra, rol** and **ror.** In the following example R1, R2, R3, and R4 are equivalent to T1, T2, T3, and T4 respectively.

```
signal S, R1, R2, R3, R4, T1, T2, T3, T4 : BIT_VECTOR (1 to 4);
...
R1 <= "sll"(S, 1);
T1 <= S(2 to 4) & '0';

R2 <= "rol"(S, 2);
T2 <= S(3 to 4) & S(1) & S(2);

R3 <= "sla"(S,2);
T3 <= S(3 to 4) & S(4) & S(4);

R4 <= S sra 1;
T4 <= S(1) & S(1 to 3);
```

It is important to note that if a numerical operand is not a globally static value, the result inferred implies more complex hardware invoking registers and multiplexers.

### 3.3.4.6. Unary operators

Unary operators "+" and "-" are defined for integer types, and they are commonly supported by synthesis tools.

```
signal R, A : INTEGER range -4 to 3;
    ...
begin
    ...
R <= -A;
```

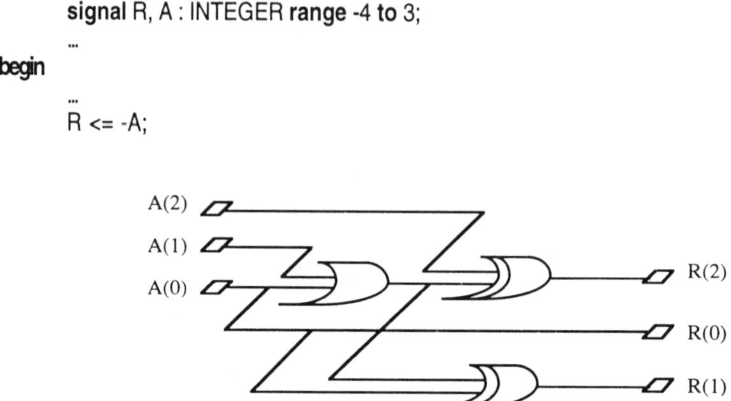


**Figure 3.10.**          **Inferred Hardware Resulting from Unary Minus Operator**

**Remark:**
Such an interpretation assumes that the position of the sign bit is defined.

### 3.3.4.7. Multiplying operators

"/" division, "*" multiplication, **mod** modulo, and **rem** remainder are considered to be multiplying operators. Only the multiplier operator is supported without restriction for all integer types. For some synthesis tools, several strategies are proposed for generating hardware. These choices can be forced on the user or left open for operating during the optimization phase.

"/" , **mod**, and **rem** are supported with restriction on values of the right operand. Synthesis tools require these values to be positive and equal to a power of two. Moreover they very often have to be global constants. The hardware result of these operators is clearly based on bit shift operators. The next example illustrates the modulo operator to build a counter modulo 4:

```
entity INCREMENTER is
port ( CLK : in BIT;
       RST : in BIT;
       R    : out NATURAL range 0 to 3);
end INCREMENTER;

architecture A of INCREMENTER is
       signal MEM : NATURAL range 0 to 3;
begin
       process (CLK)
       begin
           if CLK='0' then      --sometimes and CLK'event is required
                                -- by synthesis tools
```

```
                    if RST = '1' then
                        MEM <= 0;
                    else
                        MEM <= (MEM +1) mod 4;
                    end if;
                end if;
            end process;
            R <= MEM;
    end A;
```

### 3.3.4.8. Miscellaneous operators

**abs**, absolute value is correctly supported for all integer values. "**", the exponentiation operator, is supported with the restriction that the left operand be a global constant whose value must be evaluated at 2.


## 3.4. SEQUENTIAL STATEMENTS

All sequential statements described in this section have to be included in the sequential statement area of a concurrent process statement. Contrary to concurrent statements, the order in which these statements are written is important.


### 3.4.1. Variable Assignment

A regular (not shared) variable is declared within a process or a subprogram declaration part. A variable assignment is introduced by the := operator, and this operation is immediately taken into account. Therefore the value of a variable is changed by the next assignment statement if the new value is different from the old one. Here are different variable assignments:

```
A := 0;                     -- Literal integer value assigned to a variable
REC.FIELD := B;             -- Partial assignment of record element
VECTOR := X"AA";            -- Global vector assignment
WBUS(I) := F(A);            -- Vector element assignment
WORD(3 to 4) := "10";       -- Slice assignment
REC := ('A', 3, "10");      -- Aggregate assignment, with positional notation
REC := (LETTER => 'C',FIELD => 3, BIT2 =>"11");-- Aggregate assignment with named notation
```

Moreover, the notion of target aggregate exists on the left part of the assignment statement where several variables can be grouped. The following example shows this possibility:

```
variable V : BIT_VECTOR (1 to 3);
variable A, B, C : BIT;
...
(A,B,C) := V;                      -- These two statements are equivalent
(2 => B, 1 =>A, 3=> C) := V;
```

There is no general rule for finding out which hardware is inferred by a variable. The context in which the variable is used, determines the hardware result.

### Remark:
A variable declared within a subprogram only exists inside this context and disappears outside the subprogram. So, if this variable has been assigned, the result is lost after execution of the **return** statement unless it has been given back using **out** parameters inside a procedure or **return** statements inside a function. However this variable can provide a memorization element if a synchronization point is raised before the subprogram end.

## 3.4.2. Signal Assignment

The targets of signal assignments, always signals, have the same variety as those of variable assignments. The effect of a signal assignment is only taken into account after a synchronization statement (**wait** statement). This has two consequences: first, a signal value does not change between two synchronization statements and second, after the synchronization statement, a signal takes the value of the last assignment. The signal assignment form accepted by synthesis tools is reduced to the simplest expression:

```
Signal <= a_waveform_with_one_item_and_without_delay_expression;
```

## 3.4.3. Synchronization Statement

In VHDL the only statement to be synchronized is the **wait** statement. A process must contain at least one wait statement. In one particular case, a sensitivity list replaces an equivalent wait statement as the last statement of the process, just before the end.

In simulation, the place of the wait statement in a process is important due to object initializations, which are significant. In synthesis, as initializations are ignored, wait statements may be placed everywhere. In synthesis two different ways of using wait statements are possible:

- Waiting for an event on a signal belonging to an explicit sensitivity list. This is the way to obtain combinational hardware. In this case, this wait statement has to be unique. The sensitivity list of the wait must include all signals that are read in the process.

```
process
begin
        wait on A, B, C, D;
        S <= A or B;
        if E then
                R <= C + D;
        else
                R <= C + 1;
        end if;
end process;
```

- Waiting for an event on a signal that is recognized as a clock signal. This clock signal has to be unique and synthesis tools detect it with the help of specific interpretation of the syntax. A clock is identified by its events and edges. Thus, if a description uses a wait statement in order to be sensitive to a rising edge clock, a rising edge event causes the process to be active. For example the following conditional wait statement implies a clock signal.

```
wait until CLK='1';     -- it is equivalent to wait until CLK='1' and CLK'event
                        -- or to wait on CLK until CLK ='1'
```

If the clock signal type is not BOOLEAN or BIT, but a multivalued type, to be consistent with the simulation, the following condition has to be added to the wait statement: CLK'LAST_VALUE='0'. This determines a legal rising edge for synthesis. The clock edge condition becomes:

```
wait until CLK='1' and CLK'last_value = '0' ;
```

Only the events from '1' to '0' and vice-versa are accepted by synthesis tools as legal clock edges.

Some synthesis tools accept multiple wait statements synchronized on the same edge of the same clock signal. This authorizes the description of finite state machines without explicitly declaring the state variables.

If the wait statement is implicit (sensitivity list), the synchronization statement is described by a conditional if statement such as:

```
if CLK='0' and CLK'event and CLK'last_value='1' then ...
```

or using the attribute 'STABLE with the default time value 0 ns, ('STABLE is a function accepting a type TIME parameter which has its default value equal to zero).

**if** CLK='0' **and not** CLK'stable **and** CLK'last_value='1' **then** ...

**Remark:**
The name of the clock signal is like any identifier. No specific semantics has to be attached to a given identifier in VHDL (excepting GUARD)

## 3.4.4. Conditional Statement

Two ways of expressing conditional statements exist: **if** and **case** statements. Very often a **case** statement is a way of guiding the synthesis toward a finite state machine. Nevertheless it is also possible to extract an equivalent finite state machine from a list of **if** statements. This will be shown in the next section.

### 3.4.4.1. If statement

An **if** statement including **elsif** and **else** branches, implies a priority that is the order of execution of these different branches. Such a description infers serial multiplexers.

```
process (S1, S2, S3, R1, R2, R3, R4)
begin
  if S1='1' then
      RESULT <= R1;
  elsif S2='0' then
      RESULT <= R2;
  elsif S3 = '1' then
      RESULT <= R3;
  else
      RESULT <= R4;
  end if;
end process;
```

**Figure 3.11. Resulting Hardware of Multiplexers**

Indeed the hardware counterpart of a simple **if** statement may be built using basic gates.

```
process (OPERATOR, X, Y)
begin
  if OPERATOR='0'
  then
    RES <= X or Y;
  else
    RES <= X and Y;
  end if;
end process;
```

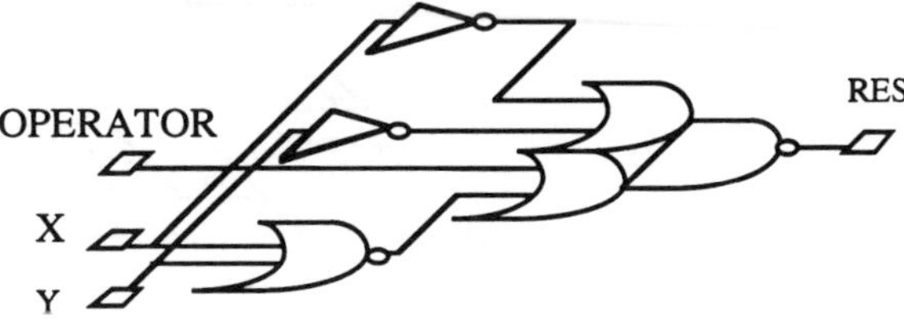


**Figure 3.12. Basic Gates Translating an "if" Statement**

### 3.4.4.2. Case statement

The use of such a statement is quite different from using an **if** statement with multiple **elsif** branches. In a **case** statement, each branch of the statement has the same precedence. In an **if** statement, **elsif** statement, each branch is sequentially checked. In a **case** statement all the possible values have to be taken into account and they are exclusive; no priority is implied. To regroup all "don't care" values, the **others** clause can be used. Note that the **others** clause should never be used if all possible values are enumerated before (*VHDL'92*). A case statement implies a multiplexed architecture.

```
type CODE_TYPE is (ADD, SUB, RST, INCX);
subtype WORD is INTEGER range 0 to 3;
signal CODE : CODE_TYPE
signal X, Y , R : WORD;

..

process (CODE, X, Y)
begin
        case X is
            when 0 => R <=Y;
            when 1 => R<= CODE_TYPE'pos(CODE);
            when others => R <= 0;
        end case;
end process;
```

Note that **others** branches are used for collecting the two and three possible values of the X signal. The next schematic diagram displays the hardware counterpart after synthesis.

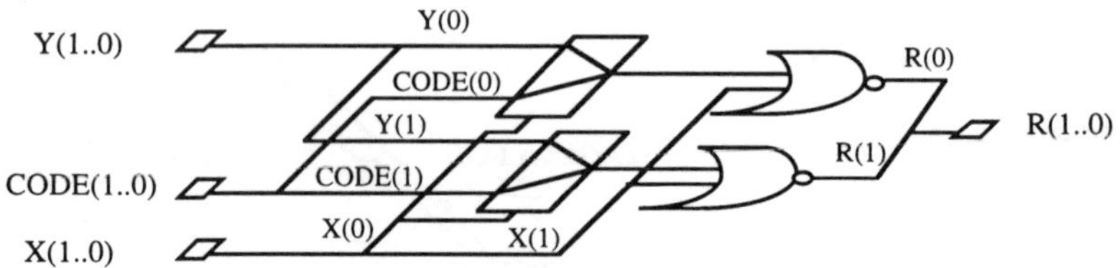


**Figure  3.13.**          **Resulting Hardware of Case Statement Using Integer**

The right lengths (2 here) of signals X, Y, and especially of signal CODE, have been computed by the synthesis tool.

The use of an enumerated type as a selector of a case statement is illustrated in the following example.

```
type CODE_TYPE is (ADD, SUB, RST, INCX);

process (CODE, X, Y)
begin
        case CODE is
            when ADD => R <= X+Y;
            when SUB => R <= X-Y;
            when RST => R <= 0;
            when INCX => R <= X+1;
        end case;
end process;
```

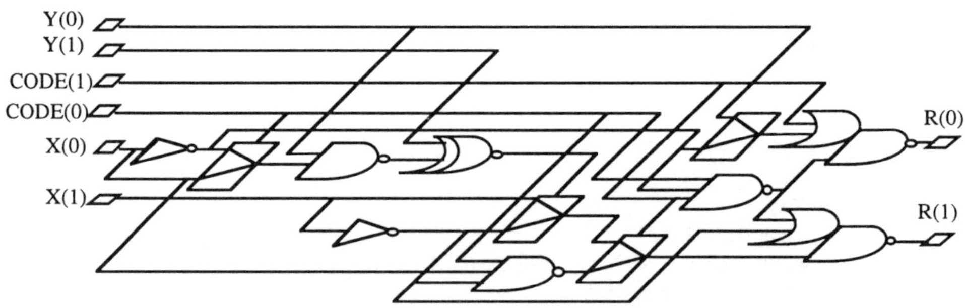


**Figure 3.14.**        **Hardware of a Case Statement Using Enumerated Type**

### 3.4.4.3. Inferring  memorization

Using conditional statements implies checking some conditions. Sometimes some condition values are not true, so the values resulting from these conditions stay unchanged. In this case, memorizations are implicit. For example, in the following **if** statement, because the CONDITION function does not always return the constant TRUE, then sometimes the RESULT will not change and will keep the same value.

```
if CONDITION(I1, I2, I3) then RESULT <= DATA(I1, I2, I3); end if;
```

Which kind of memorization is inferred here? This depends on the synchronization expression. If all read signals are in the sensitivity list, a latch is inferred. A memorization is also implied if an output signal does not appear in any possible branches of the program.

```
if CONDITION(I1, I2, I3) then RESULT <= DATA(I1, I2, I3, RESULT); end if;
```

In this last example, the RESULT is a signal that is read (right side of the signal assignment) and written (left side of the signal assignment). Thus, a memorization is needed. If this signal belongs to the sensitivity list, only a latch is inferred. On the other hand, if only the values of the inputs (I1, I2, I3) triggered by the condition, are relevant, then a flip-flop memorization is generated. Some examples below illustrate these notions.

```
process (I1, I2, I3)      -- Inferring a flip-flop
begin
        if CONDITION(I1, I2, I3)
            then RESULT <= DATA(I1, I2, I3, RESULT);
        end if;
end;

process (I1, I2, I3, RESULT)    -- Inferring a latch
begin
        if CONDITION(I1, I2, I3)
            then RESULT <= DATA(I1, I2, I3, RESULT);
        end if;
end;
```

To be sure not to infer a memorization in this last process, a default output signal assignment has to be written before the conditional statement, such as:

```
...
RESULT <= DEFAULT_VALUE;   -- Default assignment
if CONDITION(I1, I2, I3) then RESULT <= DATA(I1, I2, I3, RESULT); end if;
...
```

This first signal assignment must not be confused with the initial value explained in the previous chapter. In fact, each time the process is activated, this signal assignment is executed. Thus the value is replaced either by the default value or by the value resulting from the DATA function, depending on the CONDITION value. This next example is exactly equivalent to the previous where the **else** branch is made explicit.

```
if CONDITION(I1, I2, I3)
        then RESULT <= DATA(I1, I2, I3, RESULT);
        else RESULT <= DEFAULT_VALUE;
end if;
```

So one way of only synthesizing combinational circuits is to always put an **else** or a **when others** clause in a conditional statement. Indeed, an assignment has to be defined within these clauses. This next example, written with a **case** statement, is equivalent to the latter:

```
...
case CONDITION(I1, I2, I3) is
        when TRUE => RESULT <= DATA(I1, I2, I3, RESULT);
        when FALSE => RESULT <= DEFAULT_VALUE;
end case;
...
```

### 3.4.4.4. Resource   sharing

Within a conditional statement, resource sharing may be performed. Let us assume the following example:

```
if COND then
        RES <= A + B;
else
        RES <= B + C;
end if ;
```

A direct synthesis may infer a simple hardware, involving two adders and one multiplexer:

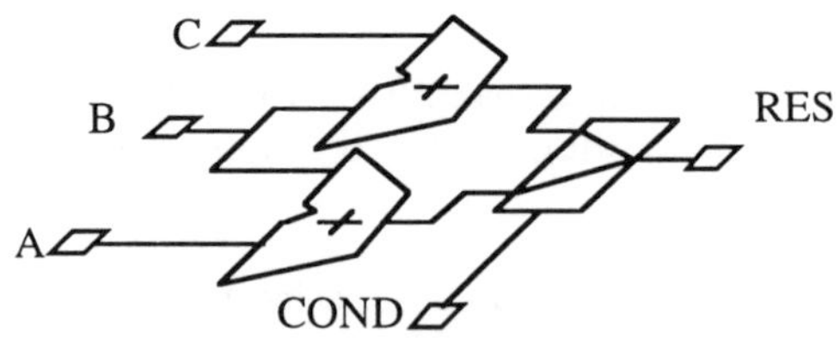


**Figure  3.15.          Resulting Hardware Without Resource Sharing**

However a more accurate synthesis may use one adder and two multiplexers because there is always only one adder used at the same time:

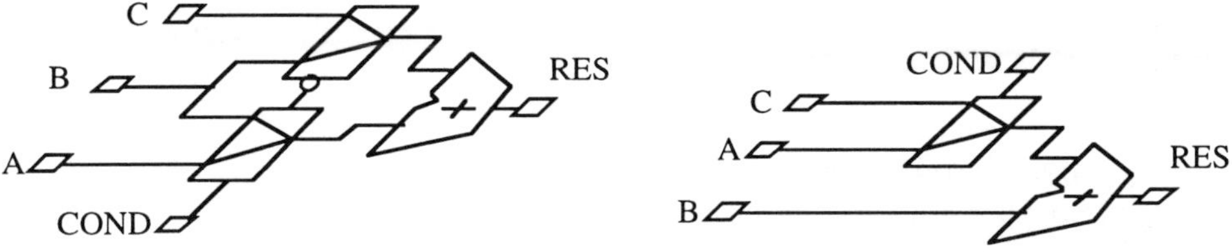


**Figure  3.16.**          **Hardware Possibilities After Adder Operator Sharing**

Resource sharing is therefore not reserved to architectural synthesis tools, and may be performed at this level of synthesis.

### 3.4.5. Iterative  Statement

In VHDL there are two kinds of iterative statements, the **for** loop and the **while** loop. The infinite **loop ... end loop** can be considered as a **while** loop with a condition that is always TRUE.

The **for** loop statement is supported by synthesis tools when the range bounds are globally static (computable at least at elaboration time).

```
signal V,S: BIT_VECTOR (1 to 10);
...
for I in V'range loop
        S(I) <= V(I);
end loop;
```

$$V(1 \text{ to } 10) \qquad S(1 \text{ to } 10)$$

**Figure 3.17. Hardware Counterpart of Assignment in for Loop Statement**

In this last example the use of the **range** attribute saves detailing the vector bounds in the loop statement. A very simple statement can be written that is totally equivalent to this **for loop** example, using the vector assignment:

```
S <= V;
```

When the range is not globally static (for example, when one of the bounds is a signal value), the synthesis result is not a simple hardware duplication. To store the loop index a register is needed, and a control must be provided to increase up to or decrease down to the register at each iteration. This kind of loop has to contain a wait statement in each possible branch to assign new values to the dynamic condition. This is the same case for the **while** statement, where the condition must be evaluated at each iteration to continue or to terminate the loop. This kind of loop is not interpreted by RTL synthesis tools, but would be the work of architectural synthesis tools.

Here is a code example whose hardware is difficult to generate by RTL synthesis tools without extra hardware.

```
...
constant N : NATURAL:= 31;
signal RG, SUM : NATURAL range 0 to N;
signal CK : BIT;
signal A : BIT_VECTOR (0 to N);
...
process
        variable CPT : NATURAL range 0 to N;
begin
        wait until CK ='1'; -- " on CK " must be added to satisfy some synthesis tools
        for J in 1 to RG loop
            if A(J) = '0' then CPT := CPT + 1; end if;
        end loop;
        SUM <= CPT;
end process;
...
```

On the other hand, the following code becomes synthesizable due to the fact that the loop bounds are constant. The hardware provided is completely flat and thus the area constraint cannot be satisfied.

There are two ways of interrupting the loop cycle before the normal end of the range: **next** and **exit** statements. For example, here is a code showing a next statement: the S vector element is assigned to the A vector element only if the corresponding bit of the COND vector is set.

```
constant COND : BIT_VECTOR(1 to 5) := "01101";
signal S, A : BIT_VECTOR(COND'range);
...
for I in COND'range loop
        next when COND(I)='0';
        S(I) <= A(I);
end loop;
```

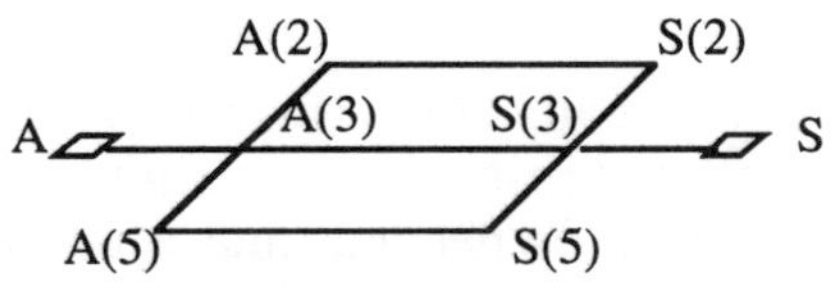


**Figure  3.18.**        **Hardware of the for Loop Statement with Next Statement**

This following part of code, counts the number of '0's contained in the beginning of vector V, starting from the index value 3 down to 0.

```vhdl
-- within architecture declarative part
signal V : BIT_VECTOR(3 downto 0);
signal SUM : NATURAL range 0 to 3;
signal CK : BIT;

...

-- within process declarative part
variable CPT : NATURAL range 0 to 3;

...

if CK = '1' and CK'event then
        for J in V'reverse_range loop
            exit when V(J) ='1';
            CPT := CPT +1;
        end loop;
end if;
SUM <= CPT;
```

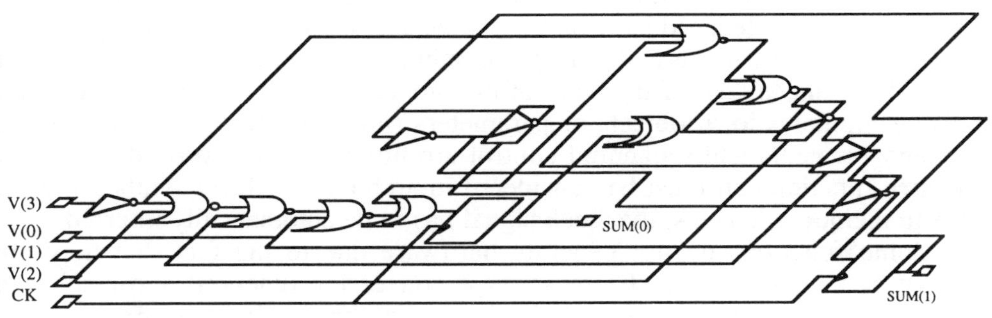


**Figure 3.19.        Synthesized Result of a Loop with Exit Statement**

An exit statement can also be used for modeling synchronous resets. The next example shows this possibility within a process that infers a finite state machine. The hardware counterpart of such a process is sequential, i.e., involves memorization elements. In this process, an infinite loop encloses a list of synchronizing statements which are all associated with checking of the RESET signal. Thus the loop is interrupted by the exit statement and the process restarts at the beginning.

```vhdl
process
begin                                   -- The signals receive
        SIGNALS <= INITIAL_VALUES;      -- their initial values
        loop
            wait until CLOCK='1';       -- Synchronizing statement

            ...
            -- Test of synchronous reset
            exit when RESET'event and RESET ='1';
```

```
            wait until CLOCK='1';
            ...
            exit when RESET'event and RESET ='1';
            wait until CLOCK='1';
            ...
            exit when RESET'event and RESET ='1';
        end loop;
    end process;
```

## 3.4.6. Subprogram   Call

A subprogram is either a procedure or a function. These are ways of
describing independent algorithms. No hierarchical level is implied by a
subprogram call. All the statements appear as if the subprogram body were flat.

A **pure** function introduced by *VHDL'92* does not cause any side effect.
Consequently the deduced hardware is always a combinational circuit.

A procedure call may also imply combinational hardware if there are **in**
and **out** parameters and if the body of the procedure, without a **wait** statement,
only manipulates local objects or parameters (no side effects). In other cases,
memory elements and sequential control are implied. A full procedure call or
**impure** function call must be considered in architectural level synthesis tools.
Due to memory elements, the clock signal has to be automatically inferred.

Some functions do not generate hardware due to the fact that they are
known by synthesis tools. These are type conversion functions and resolution
functions. These functions are only interpreted at the simulation step.


## 3.5. CONCURRENT  STATEMENTS

All concurrent statements described in this part have to be included in the
statement area of an architecture, a block statement, or a generate statement.

### 3.5.1. Process  and  Synchronization  Statement

#### 3.5.1.1. From  process  statement  to  hardware

Process statements are the most powerful VHDL constructs for modeling
behavior. They are also the basic model of every concurrent statement. When
writing such a statement, it is essential to have clear idea of its hardware
counterpart.

Does this process statement imply memory elements? Does it mean synchronous or combinational hardware? Such questions may be answered by looking at certain features of the process statement.

A process statement consists of three parts: an optional sensitivity list, a local declarative part, and a sequential statement part. Each of them has to be taken into account when establishing the characteristic (synchronous or combinational hardware) of the process.

### 3.5.1.2. Sensitivity list

A process with an explicit sensitivity list is equivalent to a process with a single synchronization statement placed just before its final reserved word **end**. The process is activated each time an event occurs on a signal of this list. Thus the process

```
process (A, B, C) -- Sensitivity list of signals
      ...
begin
      ...
end process;
```

is equivalent to

```
process
      ...
begin
      ...
      wait on A, B, C;
end process;
```

This equivalence shows that the equivalent process has a single synchronization point. Therefore, when activating this process, there is no need to "remember" at which synchronization point it was stopped. Such behavior does not imply memorization. On the contrary, *processes with multiple synchronization points, i.e., several wait statements, infer memorization.*

This is also true if the sensitivity list of the synchronization statements are identical.

```
process
begin -- A two-state finite state machine
      wait until CLK='1';
      S <= '0';
      wait until CLK='1';
      S <= '1';
end process;
```

Of course, the previous assertion does not mean that all processes with an explicit sensitivity list, and thus a single synchronization statement, are combinational: other criteria have to be taken into account.

**Remark:**
Synchronous hardware means "clock." Clock recognition, especially in the case of multiple synchronization points, is not always straightforward for synthesis tools as previously discussed in 3.4.3.

### 3.5.1.3. Local declarations

Of all possible declarations within a process declarative part, variable declarations are essential: they may imply memorization. Indeed, variables may be used in two different ways:
   • As "working" variables, they do not infer any hardware:

```
process (A, B, C) -- No memorization for this ... and gate.
        variable VAR  : BIT;
begin
        VAR := B and C;         -- The variable is the target of
                                -- an assignment before being read
        S <= A and VAR;
end process;
```

In such cases, writing the same behavior is always possible by replacing the variable with the right part of its assignment statement:

```
process (A, B, C)
begin
        S <= A and B and C;
end process;
```

Such use of variables is accepted by synthesis tools and has no hardware counterpart. In the general case:

*Between two synchronization statements, if a variable is always the target of an assignment before being read, this variable does not infer any hardware.*

   • When storing data from one process activation to another, variables obviously imply memory elements for their corresponding hardware parts.

```
process -- Synchronous finite state machine with two states
        type T_STATE is (STOP, GO);
        variable STATE : T_STATE;
```

```
begin
        wait until CLK='1';
        case STATE is              -- Variable STATE is read
                                   -- before being target of assignment
            when STOP => STATE := GO;
            when GO    => STATE := STOP;
        end case;
end process;
```

This example shows that reading a variable does not necessarily mean having it in the right part of an assignment. A **Case** expression, **if** condition or **in** parameter of a procedure or function are other ways of reading it.

*Between two synchronization statements, if at least one variable is at least once read before appearing as a target of an assignment, this variable infers memorization.*

### 3.5.1.4. Statements

Even respecting the previous two rules, i.e., a single synchronization point and variables or signals as targets of assignments before being read, a process may infer memory:

For example, this flip-flop is sequential hardware: the memorization element inferred is the flip-flop itself.

```
process
begin
        wait until CLK='1'; -- equivalent to:
                        -    wait until CLK='1' and CLK'event;

        Q <=D;
end process;
```

The notion of memorization involved here comes from the fact that between two rising edges of CLK, the value of Q is maintained, even if the value of D changes. The new value of D will only be taken into account on Q at the next rising edge of CLK.

Taking again the previous example of a two-state finite state machine, the use of a signal local to the architecture instead of a variable local to the process does not change the sequential nature of the inferred hardware: it involves memorization elements.

```
architecture A of ENT is
        type T_STATE is (STOP, GO);
        signal STATE : T_STATE;
begin
```

```
process  -- Synchronous finite state machine with two states
begin
      wait until CLK='1';
      case STATE is    -- Signal STATE is read...
                       -- but is not part of the sensitivity list
      when STOP => STATE <= GO;
      when GO    => STATE <= STOP;
      end case;
end process;
```

We have therefore to add a new rule:

*In a given process statement, if all signals whose values are read are not part of the sensitivity list, then the inferred hardware is sequential, i.e., implies memorization elements.*

The following example respects all the previous rules: a single synchronization, no variables and all read signals are part of the sensitivity list. Nevertheless, this latch obviously infers memory elements:

```
process
begin
      wait on D, G;
      if G='1' then
          Q <= D;
      end if;
end process;
```

The notion of memorization in this example comes from the fact that the output signal (Q) is not systematically the target of a signal assignment. When an event occurs on D or G and that G is equal to '0', the previous value of Q has to be memorized. In the general case:

*In a given process statement, if at least a signal is not target of signal assignment in each branch of if and case statements, then the inferred hardware is sequential, i.e., implies memorization elements.*

This last example can be completed as follows:

```
process
begin
      wait on D, G;
      if G='1' then
          Q <= D;
      end if;
      Q1 <= not Q;
end process;
```

Every event occurring on Q creates an event on Q1, and synthesis tools should not infer a second latch, but only a simple gate. Moreover this gate should be included in the latch basic cell if the complementary output is provided.

### 3.5.1.5. Summary

As a conclusion to the previous paragraphs, a process is combinational and therefore does not infer any memorization element if the following conditions are true:

- The process has its own explicit sensitivity list or contains a single synchronization point (wait statement).
- The process does not contain any variable declaration or its local variables are systematically targets of an assignment before being read.
- All signals whose values are read are part of the sensitivity list.
- All signals which are the outputs of the process are targets of signal assignments whatever the branch of the process.

Moreover, if this set of conditions is not respected, the result should infer memorization. However, only the synchronous style (clock is explained by synchronizing statements) is well supported by synthesis tools. For example, this next example, which should infer an RS memory point, is rarely supported.

```
process (SET, RESET)

        variable MEM : STD_ULOGIC;
        variable SET_RESET : STD_ULOGIC_VECTOR (1 to 2);

begin

        SET_RESET := SET & RESET;
        case SET_RESET is
            when "00" => null;
            when "10" => MEM := '1';
            when "01" => MEM := '0';
            when others => MEM := '-';
        end case;
        OUTPUT <= MEM;
        OUTPUT_BAR <= not MEM;

end process;
```

All these rules are illustrated in appendix 8.2.

## 3.5.2. Signal   Assignment

### 3.5.2.1. Simple   signal   assignment

The simplest form of a concurrent signal assignment defines a targeted signal receiving the values of the source signal each time an event occurs on the latter:

```
S <= A;          -- S is hard-wired to A;
Z <= '1';        -- In that case the signal Z is constant
```

Although combinational hardware seems to be systematically inferred by such unconditional signal statement, it is nevertheless possible to generate feedback loops in the hardware, but this is not at all recommended. A simple example of such a memory point is shown in the next code:

```
DATA <= R nand NDATA;
NDATA <= S nand DATA;
```

Several synthesis tools interpret this code as combinational and not as explicit asynchronous hardware. Process statement usage with synchronizing statements is strongly recommended to infer synchronous memories.

### 3.5.2.2. Conditional  and  selected  signal  assignments

Two concurrent signal assignments are defined for conditionally targeting a signal: conditional and select assignment. The best way to have an idea of the hardware results is to go back to the equivalent processes of these statements. For example, considering the conditional statement with conditional waveforms:

```
S <= A when X='1' else B when Y='1' else C;
```

If A, B, C, X, Y are defined as signals, the equivalent process is

```
process (A, B, C, X, Y)
begin
     if X = '1' then
          S<= A;
     elsif Y = '1' then
          S <= B;
     else
          S<= C;
     end if;
end process;
```

This process clearly respects all the conditions for inferring combinational hardware. On the other hand, this following example infers sequential hardware as certain conditions leave the signal S unchanged.

```
S <= A when X='1' else B when Y='1' else S;
```

If this statement is translated to the equivalent process, it is obvious that the signal S is read before being modified even if it keeps the same value, thus a latch and multiplexers are synthesized.

```
-- equivalent process
process (S, X, Y, A, B)
begin
        if X='1' then
            S <= A;
        elsif  Y='1' then
            S <= B;
        else
            S <= S;
        end if;
end process;
```

**Figure 3.20. Hardware Counterpart of this Conditional Assigment**

The select assignment can be translated to a process using a case statement. The same remarks as for the conditional statement can be made concerning the hardware inferred (combinational or sequential). For example, the first of the following codes infers combinational hardware, whereas the second code provides a sequential circuit.

```
-- first code
with SEL1 select
        S <=      A and B when '1' else
                  C and B when '0';
```

```
-- second code
with SEL2 select
        S <=      '0' when "00" else
                  '1' when "11" else
                  S  when others;
```

This next example, using the new key word **unaffected** introduced by *VHDL'92*, creates a multiplexer and a latch. The key word **unaffected** can be interpreted by synthesis tools without difficulty. It creates memory elements in the same way as repeating the signal on the right part of the assignment statement. To improve accuracy, an unchanged key word has been introduced to eliminate useless transactions and to improve simulation runtime. Thus both following statements are interpreted in the same way by synthesis tools.

S <= A **when** X="01" **else** B **when** X ="10" **else unaffected** ;  -- VHDL'92

S <= A **when** X="01" **else** B **when** X ="10" **else** S;                -- VHDL'87 or VHDL'92

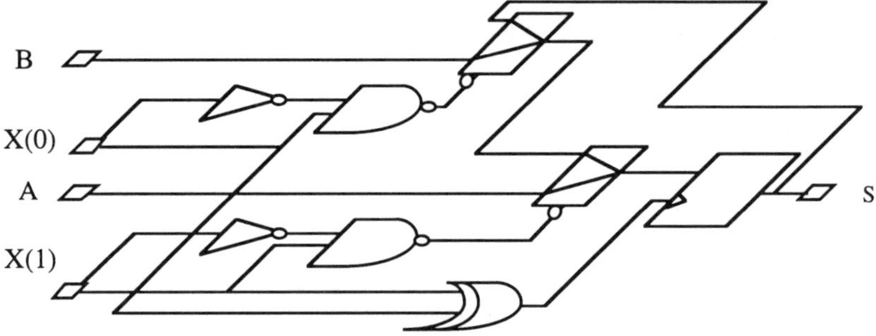


**Figure  3.21.**       **Hardware  Counterpart  of  these  Selected  Assignments**

### 3.5.3. Component  Instantiation

The purpose of a component instantiation statement is to use a model previously described. This is the main way to build the hierarchy of a design. Before an instantiation, the component has to be declared either locally within the architecture declaration part or separately in a package. The declaration of a component defines the view to be locally wired, which may thus be less general than the entity declaration. In this case, during the configuration phase some signals can explicitly be left open if they are **out** or **inout** mode, or omitted if they are **in** mode with a default value.

During the synthesis process, component instantiations can be interpreted differently. Using specific and proprietary commands of the synthesis tool, a component instantiation may be
- flattened, and thus the notion of hierarchy totally disappears
- synthesized once, and the border only of each instantiation is synthesized again depending on the context for each of them (driving value and capacity load)
- synthesized once, frozen, and duplicated for all instantiations

When instantiating a component, generic parameters have to be mapped to values and ports connected to signals.

```vhdl
architecture SYNTHESIZABLE of DRIVER is
       signal ORDER : ORDER_TYPE;
       signal CONTROL : BIT;
       signal DATA_IN, DATA_OUT : BIT_VECTOR (1 to N);
       component CTRL
           port (C : in BIT; O : out BIT_VECTOR);
       end component;
       component OPRT
           port (O   : in BIT_VECTOR;
                 DI  : in BIT_VECTOR;
                 DO : out BIT_VECTOR);
       end component;
       -- Specification configuration
       for CONTROLER : CTRL use entity WORK.CTRL(A);
       for OPERATIVE_PART : OPRT use entity WORK.OPRT(A) generic map (N);
begin
       -- Two component instantiation statements
       CONTROLER : CTRL port map (CONTROL, ORDER);
       OPERATIVE_PART : OPRT port map (ORDER, DATA_IN, DATA_OUT);
end SYNTHESIZABLE;
```

The configuration specifies which pair entity/architecture is chosen for
component instantiation. In the last example, the two configuration specification
statements may be replaced by one stand-alone configuration:

```vhdl
configuration C1 of DRIVER is
       for  SYNTHESIZABLE
            for CONTROLER : CTRL use entity WORK.CTRL(A); end for;
            for OPERATIVE_PART : OPRT use entity WORK.OPRT(A)
                generic map (N);
            end for;
       end for;
end C1;
```

Configurations are not very often supported by synthesis tools.
Nevertheless there is no real difficulty in implementing them. They have to be
recognized in order to promote compatibility between the simulation and
synthesis environments.


## 3.5.4. Block  Statement

Simulation block statements are used for structuring concurrent statements.
The simple way is logical grouping for readability. This use has no effect on
the synthesis result. The following example illustrates local declarations:

```vhdl
signal A, B, C, D, E, F, G : BIT;
signal S : BIT_VECTOR(1 to 4);
```

```
...
BLOCK_NAME : block
        subtype TWO_BIT_TYPE is BIT_VECTOR (1 to 2);
        signal V1,V2 : TWO_BIT_TYPE;
begin
        V1 <= (A or C ) & B;  V2 <= E & (F and G);
        S <= V1 & V2;
end block BLOCK_NAME;
```

**Remark:**

Under certain conditions, some synthesis tools can interpret the block statement as a hierarchical level of synthesis. In this case the label of the block statement is used as a parameter for structuring commands.

Another use of block statements is to define a concurrent statement part controlled by a clock signal. In this case a guard section is required to describe explicitly the clock signal and its associated condition (rising or falling edge). Moreover, concurrent signal assignments have to be guarded to be activated by the guard expression. This is shown in the next example where it is assumed that the type of signal CLK is multivalued.

```
B : block (not CLK'stable and CLK ='0' and CLK'last_value ='1')
        signal R : BIT;
begin
        R <= guarded DATA(1);
        S <= guarded R;          -- Signal S is DATA (1) delayed for two cycles
end block;
```

The third use of a block statement is devoted to three-state circuitry. Associated with the disconnect clause and resolved signal, a guarded assignment can define three-state behavior. When the guard condition is false, the targeted signal is disconnected at the same time as the guard condition becomes false. This is the default behavior of the disconnect clause. Such a disconnection is authorized if the type of targeted signal is resolved. For synthesis, the problem to be solved is the final value when all the source signals are disconnected. This value depends on two elements: the result of the resolution function returns (when the length of the input vector is null) and the kind of resolved signal (**bus** or **register**). To summarize the behavior of this kind of signal

- the final result is the value returned by the resolution function with a null input vector if the type of targeted signal is **bus**
- the final result is the previous value when there is only one source signal and if the type of targeted signal is **register**.

Certain synthesis tools support the disconnected signal of kind bus, if the result of the resolution function explicitly returns the 'Z' coding the three-state behavior. There is no major difficulty in implementing a pull-up or pull-down: the resolution function respectively returns 'H' or 'L'.

Synthesis of a resolved signal of kind register is more delicate. In this case a memory should be inferred, and the difficulty is to compute the logic circuitry to control the memory element. This is usually left to architectural synthesis tools.

This next example shows three-state buffers inferred by implicit disconnection of the source signal within a block statement with a guarded condition.

```
use IEEE.STD_LOGIC_1164.all
signal ENABLE : STD_ULOGIC;
signal DATA : STD_ULOGIC_VECTOR(1 to 2);
signal S : STD_LOGIC_VECTOR (1 to 2) bus ;

...
TRI_STATE : block (ENABLE = '1')
      begin
            S <= guarded  STD_LOGIC_VECTOR(DATA);
      end block TRI_STATE;
```

**Remark:**
DATA and S are not of the same type, and therefore the type of S is a vector of resolved STD_LOGIC. However the bit elements of S are of the same base type of DATA, which is the reason why a type conversion must be carried out between DATA and S during the signal assignment.

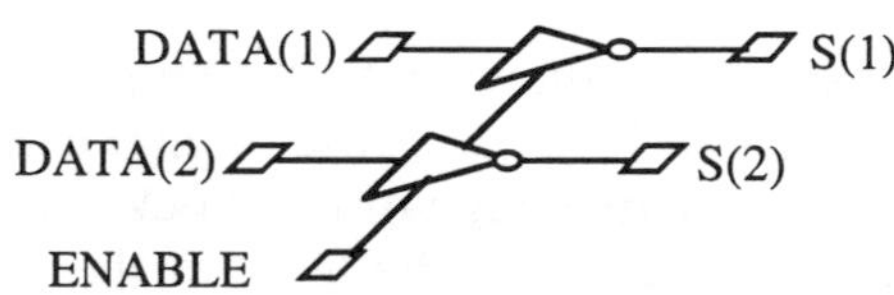


**Figure  3.22.**          **Schematic of Three-state Buffer Inferred by Block Statement**

## 3.5.5. Concurrent  Procedure  Call

A concurrent procedure is a procedure called from within a block or an architecture. As explained in the previous chapter, a process exists which is equivalent to the concurrent procedure call. The process equivalent to the concurrent procedure call is shown in the next code, using a wait statement and sequential procedure call:

```
        P(A,B,C);      -- concurrent procedure call;
```

is equivalent to:

```
        process          -- process equivalent to the previous concurrent procedure call,
        begin            -- C is assumed to be an actual parameter of mode out.
            P(A,B,C);
            wait on A,B;
        end process;
```

**Remark:**
The LRM authorizes a **wait** statement within concurrent procedures. In the synthesis domain, this not supported. Indeed multiple wait statements are only authorized if the synchronizing conditions are the same.

Within the body of a concurrent procedure a signal can be read or targeted even if it is not a parameter (side effect). This possibility reduces the general use of procedures. The consequences on the synthesis domain are that when the signal is targeted from different procedure calls, it must be resolved and that when it is a source signal, the only possible signal as parameter of the procedure is a clock signal.

A common use of concurrent procedure calls is when flexibility and adaptability are targeted. A concurrent procedure call is easier to use than a component instantiation which requires declaration and specification. Nevertheless concurrent procedures do not offer all the possibilities provided by instantiation, and local signal declarations are especially forbidden in a procedure.

The next example shows a concurrent procedure modeling a parametrized D flip-flop register. The active level of a reset signal can be changed, which by default is the high level. The active edge of the clock can be specified and this is the rising edge by default. And finally the level of the output is flexible, which by default is the direct output.

The procedure body is made up of two parts: the asynchronous part to treat the reset signal and the synchronous part to compute the next data output.

```
procedure DFF ( signal D      : in BIT;
                signal CLK    : in BIT;
                signal RST    : in BIT;
                signal Q      : out BIT;
            constant RST_LEVEL : BIT := '1';
            constant CLK_EDGE  : BIT := '1';
            constant Q_LEVEL   : BIT := '1') is
begin
        if RST = RST_LEVEL then
            Q <= not Q_LEVEL;
```

```
            elsif (CLK'event and CLK=CLK_EDGE) then
                case Q_LEVEL is
                    when '1' => Q <= D;
                    when '0' => Q <= not (D);
                end case;
            end if;
    end DFF;
```

## 3.5.6. Generate Statement

Two generate statements are defined by the LRM: the iterative and the conditional generate statements. From the synthesis point of view, these kinds of statements are not always fully supported. During the elaboration phase, a set of equivalent concurrent statements has to be generated.

### 3.5.6.1. Iterative generate statement

This statement enables repetition of a set of concurrent statements (for example, to repeat several components or to assign vectored signals).

```
architecture A of E is
        constant k1    : NATURAL := 0;
        constant k2    : NATURAL := 3;
        signal A,S     : BIT_VECTOR(k1 to k2);
        signal R       : BIT_VECTOR(k1+1 to k2+1);
begin
        L : for J in k1 to k2 generate
            S(J) <= not A(J);
            R(J+1) <= A(J);
        end generate;
end ;
```

**Remark:**
The constants K1 and K2 could be generic parameters and the hardware counterpart will not change, as explained in the next section.

In the following example, architectures A and B are equivalent, and the final synthesis results are identical. Within the first architecture A, the vector result V is assigned within a sequential loop statement, and within the second architecture B, N concurrent signal assignments are generated.

```
architecture A of E is
        constant N : NATURAL := 3;
        signal B : BIT_VECTOR (0 to N);
        signal A,V : BIT_VECTOR (1 to N);
begin
```

```
        process (A, B)
        begin
            for I in 1 to N loop
                V(I) <= F(I, A(I), B(N-I + 1));
            end loop;
        end process;
end A;
architecture B of E is
        constant N : NATURAL := 3;
        signal B : BIT_VECTOR (0 to N);
        signal A,V : BIT_VECTOR (1 to N);
begin
        L : for I in 1 to N generate
            V(I) <= F(I, A(I), B(N-I + 1));
        end generate;
end B;
```

### 3.5.6.2. Conditional generate statement

To complete a modeling using iterative generate statements, it can be useful to check the first and last values of the loop index. Conditional generate statements can be used to solve this problem. The following example illustrates this property, where the treatment of component instances may be different (none, one or several instances).

```
entity E is
        generic (N : NATURAL);
        port (INPUT : in BIT; OUTPUT : out BIT; ...);
end E;
architecture A of E is
        signal LOCAL : BIT_VECTOR (1 to N-1);
        component M1
            port (I : in BIT; O : out BIT);
        end component;
begin
        L1 : if N = 0 generate
            OUTPUT <= INPUT;                          -- no component instantiation
        end generate;
        L2: if N = 1 generate
            C1 : M1 port map (INPUT, OUTPUT);      -- only one instantiation
        end generate;
        L3 : if N >= 2 generate  -- other cases using the intermediate LOCAL signal
            F : M1 port map (INPUT, LOCAL(1));
            L4 : for I in 1 to N-2 generate
                C : M1 port map (LOCAL(I), LOCAL(I+1));
            end generate;
            L : M1 port map (LOCAL(N-1), OUTPUT);
        end generate;
end A;
```

## 3.6. USING  GENERICS

A generic parameter is a generalization of a constant notion, that is to say, during the elaboration phase, their values must be evaluated. So before this step, for the compilation phase, their values are only typed but unknown. Some synthesis tools impose restrictions on generic parameter types (only integer and enumerated types are supported). However, there is no real reason for such restrictions, except for not supporting the configuration. Indeed the final values of the generic parameter are given either by a configuration specification statement or within a configuration unit.

The first example shows how to write a model, which accepts a vector with a size of N, as input. N is effectively the generic parameter of this model.

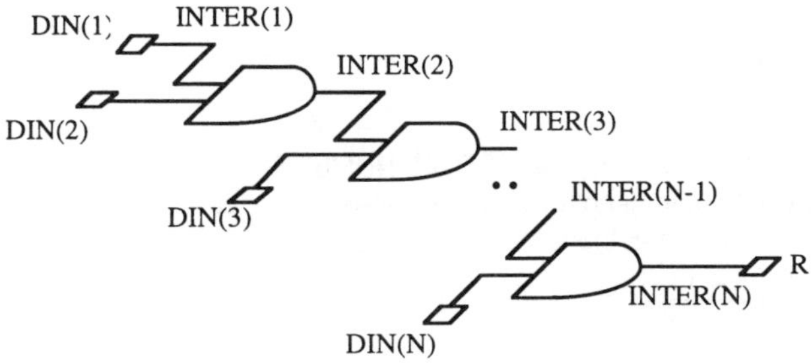


**Figure  3.23.**          **Generic N Inputs and Gate**

```
entity AND_N is
      generic ( N    : POSITIVE );
      port (    DIN : in BIT_VECTOR(1 to N);
                R    : out BIT);
end AND_N;
```

```
architecture A1 of AND_N is                    architecture A2 of AND_N is
      signal INTER  : BIT_VECTOR(1 to N)           begin
begin                                                  process (DIN)
      INTER (1) <= DIN(1);                                 variable RES : BIT;
      L: for I in 1 to N-1 generate                begin
            INTER (I+1) <= DIN(I+1)                        RES := DIN(1);
                        and INTER(I);                      for I in 2 to N loop
      end generate;                                           RES := RES
      R <= INTER(N);                                              and DIN(I);
end A1;                                                     end loop;
                                                           R <= RES;
                                                     end process;
                                               end A2;
```

**Remark:**
This example can be easily generalized to all associative functions
(f(x,(y,z)) = f((x,y),z)), for example: or, addition, multiplication and also serial
register, ...

The second example shows a model that recognizes a given pattern in a
serial bit flow. The generic parameters are the vector to be found, and also the
number of bits authorized as error number (0: the pattern must be found
exactly as described, 1: only one bit difference can be accepted, ...)

```vhdl
entity SEARCH_PATTERN is
        generic ( PATTERN          : BIT_VECTOR;
                  ERROR_NUMBER : NATURAL);
        port (    CLK      : in BIT;
                  DATA     : in BIT;
                  RESET    : in BIT ;
                  FOUND    : out BIT);
begin
        assert (PATTERN 'length >= ERROR_NUMBER)
            report "Pattern length cannot be shorter than " &
                      " the authorized error number"
            severity ERROR;
end SEARCH_PATTERN;

architecture A of SEARCH_PATTERN is
        subtype REG_TYPE is BIT_VECTOR(PATTERN'range);
        signal REG : REG_TYPE;
begin
        P_REG : process (DATA, CLK, RESET)
        begin
            if RESET ='1' then
                REG <= REG_TYPE'(others => '0');
            elsif CLK = '1' and CLK'event then
                REG <= REG(REG'left-1 downto REG'right) & DATA;
            end if;
        end process;

        P_FOUND : process (REG)
            variable CPT : NATURAL range 0 to ERROR_NUMBER +1;
        begin
            CPT:= 0;
            FOUND <= '0';
            for I in REG'range loop
                if PATTERN(I) /= REG (I) then
                    CPT := CPT + 1;
                    if CPT >= ERROR_NUMBER then
                        FOUND <= '1';
                        exit;
                    end if;
```

```
                end if;
              end loop;
          end process;
  end A;
```

## Remarks:

- The use of assert statements within the entity must not infer hardware. This statement checks consistencies between parameter values that are difficult to describe by static expressions (such as ranges in type or subtype expressions).
- The architecture is composed of two processes: the first infers sequential hardware with an asynchronous reset, whereas the second implies purely combinational circuitry.
- The REG_TYPE declaration has been introduced to qualify the aggregate (others => '0') in the asynchronous RESET branch.
- Generic entities need to be mapped to parameter values to produce real hardware. This can be done during component instantiation or during configuration where generic parameters are mapped to their values. The following code shows two different architectures. Architecture A declares a nongeneric component and specifies the parameter values of the generic entity which is configurated. The component declared within architecture B is generic and its parameter values are mapped at component instantiation. Figure 3.24 shows the differences in during configuration. These two approaches lead to the same inferred hardware.

```
entity PATTERN is
      port (DATA   : in BIT;
            CLK    : in BIT;
            RESET  : in BIT;
            FOUND  : out BIT);
end PATTERN;

architecture A of PATTERN is
      component P2_0
            port (CLK    : in BIT;
                  DATA   : in BIT;
                  RESET  : in BIT;
                  FOUND  : out BIT);
      end component;
-- Specification of configuration mapping generic parameters
      for C1 : P2_0 use entity WORK.SEARCH_PATTERN(A)
                  generic map (PATTERN => "11", ERROR_NUMBER=> 0)
                  port map (CLK, DATA, RESET, FOUND);
begin
      C1 : P2_0 port map (CLK, DATA, RESET, FOUND);
end;
```

```
architecture B of PATTERN is
        component P7_2
            generic ( PATTERN          : BIT_VECTOR (6 downto 0);
                      ERROR_NUMBER : NATURAL);
            port (CLK     : in BIT;
                  DATA    : in BIT;
                  RESET   : in BIT;
                  FOUND   : out BIT);
        end component;
    -- Using configuration specification by default
        for C1 : P7_2 use entity WORK.SEARCH_PATTERN(A);
begin
-- Component instantiation mapping generic parameters
        C1 : P7_2     generic map (PATTERN =>"0000000", ERROR_NUMBER=> 2)
                      port map (CLK, DATA, RESET, FOUND);
end;
```

Figure 3.24 shows the inferred hardware (corresponding to either architecture A or architecture B) with two different sets of parameter values.

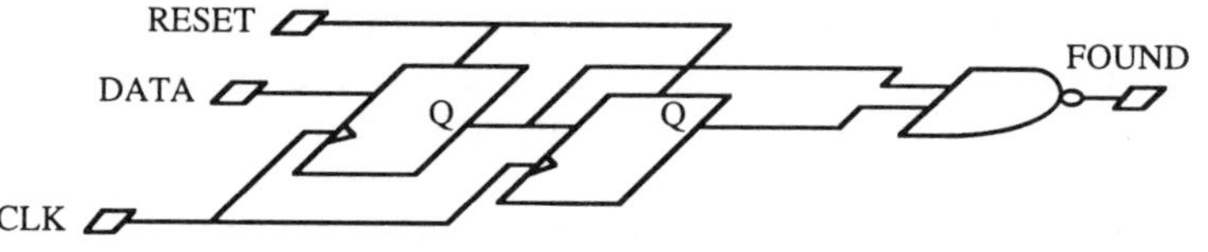


**Generic parameters mapping to ERROR_NUMBER=0 and PATTERN="11"**

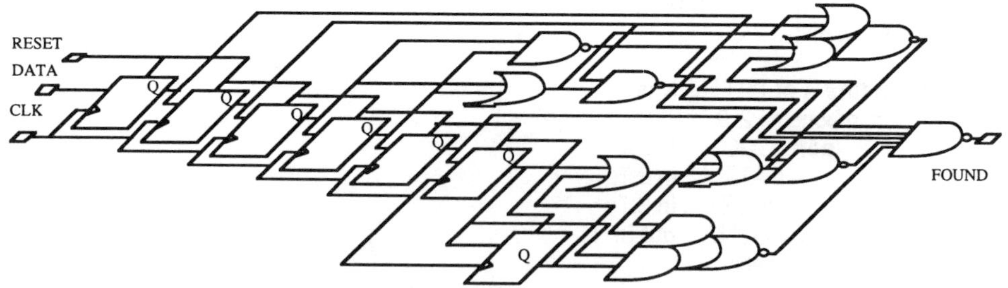


**Generic parameters mapping to ERROR_NUMBER=2 and PATTERN="0000000"**

**Figure 3.24.          Resulting Hardware of Two Different Instantiations**

## 3.7. CONCLUSION

In this chapter, a synthesis semantics has been given to a subset of VHDL constructs. These restrictions to the full language features have different reasons:
- the lack of maturity of synthesis tools such as the deficiency of elaboration phase and its consequences on configurations, generic parameters, and generate statements
- the state-of-the-art in synthesis only targets RTL synthesis
- certain VHDL features are simply not synthesizable: VHDL is a language which has been designed with a simulation semantics.

Nevertheless, an efficient design may be obtained using synthesis if VHDL code has been written without forgetting the hardware implied. In the next chapter, typical hardware examples are given. Their descriptions follow the rules and restrictions described in this chapter.

# 4. MAPPING HARDWARE TO VHDL

Throughout the previous chapter, VHDL constructs have been studied to extract their usable subset from the point of view of synthesis. VHDL semantics for synthesis is now more clear. An accurate and nonambiguous semantics is essential for obtaining a result reasonably independent of the description style. Here is the most important issue concerning the generality of VHDL: the generation of hardware has to be as independent as possible of the modeling style.

After presenting the essential vocabulary, this chapter shows different modeling styles. Its goal is to offer a methodological guide for choosing a description targeting a given hardware resource.

Synthesizable level modeling is one of the ultimate steps of the design cycle. It allows an adequate abstraction level to avoid an explicit description of the physical resources. For optimal efficiency of the synthesis process, the designer must first know how to accurately model classical hardware resources. Once this knowledge is acquired, this chapter will show how to target more complex descriptions.

The following approach is taken. First, the know-how of synthesis tools in combinational, and sequential logic is studied. Then, each of these two families is explored to point out their basic descriptions. To benefit more from the power of VHDL in abstractions, complex examples are then taken. Therefore, certain of these descriptions are so abstract (finite state machines for example) that it will not be possible to (a priori) determine the exact architecture.

## 4.1. COMBINATIONAL CIRCUITS

As mentioned in chapter 3, a block is said to be combinational if it does not imply any memorization: its outputs are only dependent on its inputs, not on an internal state.

The most popular combinational devices are the elementary logic gates: for all possible values of the inputs, the output value may be anticipated. In this section, the following elementary hardware resources are studied:
- simple logic gates
- comparators
- arithmetic operators
- shift and rotate operators
- multiplexers

In a second step, more complex devices are examined. Indeed, they are made up of the previous elements. The following devices are considered:
- simple and more complex ALUs
- three state logic (bidirectional tristate amplifiers)
- ROMs
- PLAs

## 4.1.1. Logic   Gates

The most appropriate modeling style for such descriptions is the concurrent signal assignment. For example, the simplest solution for describing a NOR gate with two inputs is the following:

```
entity NOR2 is
        port (    A, B     : in  BIT;
                  S        : out BIT);
end NOR2;

architecture COMB of NOR2 is
begin
        S <= A nor B;
end COMB;
```

Of course, other styles exist and although longer, they may sometimes be more readable:

```
architecture COMB1 of NOR2 is
begin
        S <=    '0' when A = '1' or B = '1'
                else '1';
end COMB1;
```

Such a description may also be expressed using an explicit process statement. The example below is indeed an equivalent process to the previous conditional assignment.

```
architecture COMB2 of NOR2 is
begin
        P_NOR2 : process (A, B)      -- All read signals are part of the sensitivity list
        begin
            if A = '1' or B = '1' then
                 S <= '0';
            else                            -- The output signal is assigned within each branch
                 S <= '1';
            end if;
        end process P_NOR2 ;
end COMB2;
```

Other data types, such as type BOOLEAN or exported types of package STS_LOGIC_1164 may also be used. Their synthesis semantics is the same. Choosing between them is a modeling issue.

Type BOOLEAN allows for a little more abstraction (no coding value) and may thus improve the general readability of the description.

However, types exported from package STD_LOGIC_1164 are closer to the electrical level. They allow more accuracy:
- logic states (states '0', '1', 'L' and 'H')
- conflict result (states 'X' and 'W')
- connection or disconnection of a signal (state 'Z')
- initialization value (state 'U' for default value before any assignment)
- "don't care" value (state '-') — This last value allows for not systematically specifying, for example when using conditional or selected statements, the value of the outputs. Obviously, this underspecification leaves the synthesis tool free to perform some optimizations.

STD_LOGIC_1164 is a standard package stored in a library called IEEE. Therefore, it is necessary to reference it (before using it) with a **library** clause. Two types exported by this package may be used for logic values:
- STD_ULOGIC is made up of nine states and is a nonresolved type: signals of this type are not allowed to have multisources
- STD_LOGIC is also composed of nine states and is a resolved subtype of the previous one: signals with multiple sources are allowed and conflicts are resolved.

Sections 3.2 and 6.2 provide more information on this package. Using type STD_ULOGIC, the previous description becomes:

```
library IEEE;
use IEEE.STD_LOGIC_1164.all;
entity NOR2 is
        port (    A, B     : in  STD_ULOGIC;
                  S        : out STD_ULOGIC);
end NOR2;
```

```
architecture COMB of NOR2 is
begin
        S <= A nor B;
end COMB;
```

The expression on the right side of the signal assignment may also be vectorized. The following example generalizes the previous hardware resource by parametrizing it: a **generic** clause is used.

```
entity NOR2 is
        generic (N : POSITIVE := 2);
        port (    A, B    : in   BIT_VECTOR(N-1 downto 0);
                  S        : out BIT_VECTOR(N-1 downto 0) );
end NOR2;
architecture COMB of NOR2 is
begin
        S <= A nor B;
end COMB;
```

Only a single elementary operation has been defined in all the previous examples. The same approach may be used for all predefined logical operators such as **and**, **nand**, **or**, **xor**, or **not**. Therefore, complex logical expressions may be written.

When the same combinational treatment is used more than once in the same design, the ability to reuse it is important. VHDL offers several solutions providing this functionality. As seen in chapter 2, the hardware hierarchy provides a general method based on the instantiation of a component and thus its configuration. Direct instantiation (cf. paragraph 2.2.2) is also a possibility but is not accepted by many tools (*VHDL'92*). Indeed, when the treatment to be reused is not complex, software hierarchy and more particularly subprogram calls may be used. The notion of (pure) function is particularly well adapted to combinational treatments: the returned value is only a function of the function input parameters. A function call is itself an expression and therefore may be used on the right side of a signal assignment. Indeed, procedures also allow for describing reusable combinational part, and examples will be given later (cf. paragraph 4.2.5). The example below uses a function for the parity computation:

```
function PARITY (BV : in STD_ULOGIC_VECTOR) return STD_ULOGIC is
        variable RESULT : STD_ULOGIC;
begin
        RESULT := '0';
        for I in BV'range loop
            RESULT := RESULT xor BV(I);
        end loop;
        return RESULT;
end PARITY;
```

The main interest of this description is based on its high degree of flexibility due to its unconstrained parameter. The context of call defines the size of this object as well as the required number of hardware resources.

Where do functions have to be declared? The answer to this question depends mostly on the characteristics of such a function: a very specific function is usually locally declared, whereas a more general one is likely to be declared in a package. The following example describes how to use the general purpose function PARITY supposed to be declared in package PKG and compiled in library WORK.

```
use WORK.PKG.all;
use IEEE.STD_LOGIC_1164.all;
architecture COMB of DESIGN is
        signal DATA : STD_ULOGIC_VECTOR(7 downto 0);
        signal P        : STD_ULOGIC;
begin
        ...
        P <= PARITY( DATA );
        ...
end COMB;
```

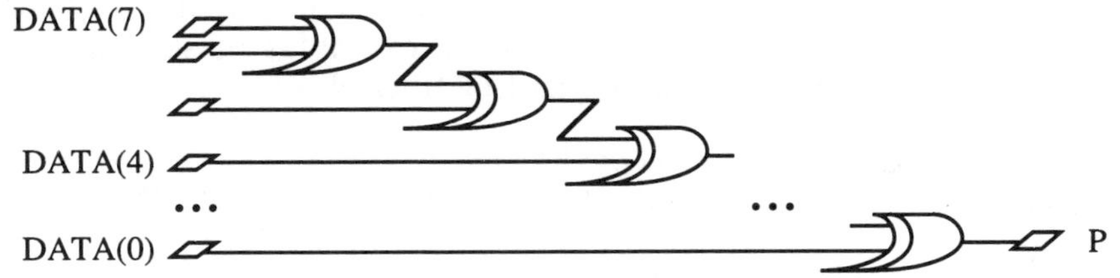

DATA(7)

DATA(4)

...

DATA(0)                                                                    P

**Figure 4.1. Parity Computation**

## 4.1.2. Comparators

When modeling, a common objective is to allow a treatment in a function of the threshold of a data or when a command is in a given state. Indeed, two kinds of comparisons may be distinguished:

- Checking that two objects are identical. Functions (also called operators) of equality ('=') and inequality ('/=') are involved. All synthesizable types (see chapter 3) may be compared using these operators.
- Checking the order of two objects. Using relational operators, these operations may be performed on any type with a related order such as integer types, enumerated types and ASCII representation types for which comparison operators have been overloaded.

**Remark:**
Checking the order of enumerated types may lead to a trap in some cases. Since the synthesis process translates enumerated into bit vectors, this operation implies a certain coding. By default, this coding is consistent with the order of the enumerated type and only involves the necessary number of bits needed. With the following declaration,

```vhdl
type STATE is (STOP, GO, PAUSE);
```

the enumerated values are coded with two bits. STOP is coded "00", GO becomes "01", etc. The order is therefore respected. The problem arises when the synthesis tool provides a way for specifying a given coding. If the designer gives "11" for the coding of STOP, there will be an inconsistency between the result of the comparison before synthesis (performed on the enumerated type) and the result of the comparison after synthesis (performed on the bit vector). Therefore, the simulations of the two descriptions may differ.

Comparisons are mainly used in selected and conditional statements of the sequential and concurrent domains. Nevertheless, they may also appear as simple signal or variable assignments: the type of the assignment target is therefore of type BOOLEAN. Here are some examples of such assignments where S is supposed to be a signal of type BOOLEAN:

```vhdl
-- SPEED is supposed to be a signal of type NATURAL:
S <=  SPEED > 273;              -- is a comparison between a signal and a constant.

-- T_INT and T_EXT are two signals of type NATURAL
S <=  T_EXT >= T_INT;          -- is a comparison between two signals.

-- B is supposed to be a signal of type BIT_VECTOR:
S <=  B = "00010110";          -- is also a comparison between a signal and a constant.

-- E is a signal of an enumerated type involving values A and D:
S <=  (E = A) or (E = D);

-- R is a variable of type BOOLEAN, VECTOR is a signal of type BIT_VECTOR and
-- ENABLE is a signal of type BOOLEAN:
process (A, ENABLE, ...)
        variable R : BOOLEAN;
begin
        ...
        R := (A = "1010") and ENABLE; -- R may be used within this process
        ...
        if R then
            ...
        end if;
end process;
```

These simple examples imply some remarks. First, the fact that this comparison involves a constant may be used by the synthesis tool for minimizing the deduced hardware (as shown in figure 4.2).

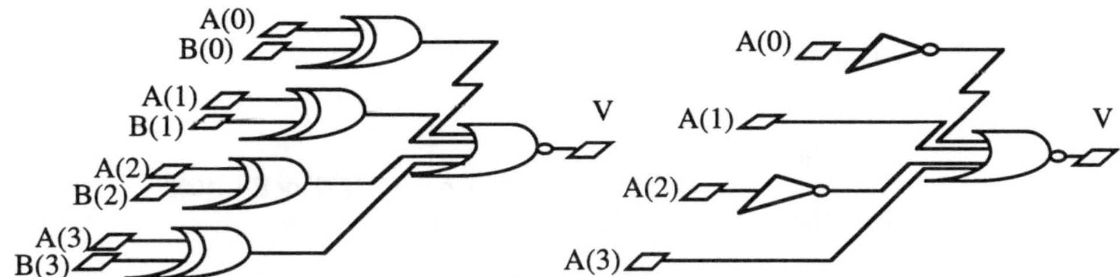

**Comparison of signals A and B**          **Comparison of signal A and constant "1010"**

**Figure  4.2. Inferred  Hardware  Minimization**

## Second remark:

These operators are predefined, and therefore there is no need to "use" a specific package.

In contrast, the following examples illustrate the use of types for which the comparison operators have to be redefined (overloaded). These types are mainly vectors of types BIT_VECTOR, STD_ULOGIC_VECTOR, STD_LOGIC_VECTOR or related types.

In the next example, signals POSITION_NEW and POSITION_OLD are supposed to be integers:

```
library IEEE;
use IEEE.STD_LOGIC_1164.all;
...
architecture A of DESIGN is
        signal POSITION_NEW, POSITION_OLD : STD_ULOGIC_VECTOR (9 downto 0);
begin
        P_CMP : process (POSITION_NEW, POSITION_OLD, ...)
        begin
                ...
                if POSITION_NEW >= POSITION_OLD then    -- A true 10-bit comparator
                        -- Ascending treatment
                else
                        -- Descending treatment
                end if;
                ...
        end process P_CMP;
end A;
```

Writing conditional assignments of variables or signals to given values is a common way to express Boolean equations. In the example below, given an input signal E of type BIT_VECTOR with a length of 4, an output signal S of type NATURAL (**range** 0 **to** 4) is computing the number of '1's in the value of signal E.

```
library IEEE;
use IEEE.NUMERIC_BIT.all;
...

...
S <=              0   when   E = "0000"
        else      1   when   E = "0001" or E = "0010" or E = "0100" or E = "1000"
        else      2   when   E = "0011" or E = "0110" or E = "1100" or
                             E = "1010" or E = "0101" or E = "1001"
        else      3   when   E = "0111" or E = "1011" or E = "1101" or E = "1110"
        else      4;
```

The equality operator used here is the base operator predefined with type BIT_VECTOR. In this case, there is no need to define another one, as the comparison operation only deals with bit vectors of the same length and therefore a predefined "=" operator on type BIT_VECTOR is appropriate.

Nevertheless, in other situations, this predefined operator may lead to a trap: bit vectors are systematically different if they have different lengths. For example, "00111" is different from "0111". Therefore, the designer has to be careful when using it.

If type STD_ULOGIC_VECTOR (or another extended logic type) were used instead of BIT_VECTOR, the use of an overloaded equality operator would be necessary. Such an operator has to be used in order to deal with comparison of values such as 'X' or 'Z'.

## 4.1.3. Arithmetic  Operations

These target, after Boolean operations and comparisons, another important family of hardware resources. This paragraph emphasizes the four basic operations: addition, subtraction, multiplication, and division.

The most convenient data type for performing such operations is obviously type INTEGER and its associated subtypes: NATURAL and POSITIVE. The VHDL operators '+', '-', '*' and, '/' are predefined and may thus be used without any explicit reference to a package.

It is essential for the designer to constrain the size of his or her objects of type INTEGER to the optimal dimension. If he or she fails in this task, the synthesis tool assumes the definition size of type INTEGER in package STD.STANDARD. This value is implementation dependent but is usually equal to 32 (as the usual proprietary definition of type INTEGER in package STD.STANDARD).

Here is a first simple example:

```
-- Somewhere in a declarative part of the concurrent domain (architecture for example).
signal I1, I2, SUM_I : INTEGER range -16 to 15;
signal N1, N2, SUM_N : INTEGER range 0 to 47;


-- Somewhere in the corresponding statement part..
SUM_I    <=  I1   +   I2;
SUM_N    <=  N1   -   N2;
```

This short example needs some comments about the encoding convention of numbers. All the synthesis process aims to translate an HDL source code into a binary representation. Here, signed numbers I1, I2, and SUM are coded in two's complement using 4+1 bits. On the other hand, unsigned numbers N1, N2, and SUM_N are coded with 6 bits ($47 < 2^{**}6-1$). Nevertheless, it is not necessary to give systematically the range of objects of type INTEGER as a power of two. For example, if P is a POSITIVE representing the size of an object, the following cumbersome declarations may be avoided:

```
signal I    : INTEGER range  -2**(P-1)      to 2**(P-1) - 1;    -- signed
signal N    : INTEGER range  0              to 2**P - 1;        -- unsigned
```

Indeed, the synthesis tool is able to deduce the convenient size of the binary representation whatever the range.

```
signal K : INTEGER range N to M;
```

If N is positive or equal to zero, an unsigned representation is chosen with a consistent length L which is the smaller number such as $(2^{**}L)-1 >= M$. If N is negative, a signed representation is necessary. Length L is the smaller number such as $(2^{**}(L-1)-1 >= M$ and $-2^{**}(L-1)<=N$.

Nevertheless, designers have to be careful; the efficiency of the synthesis result depends on the accurate control they can provide of the hardware resources used and therefore on their optimal size.

Arithmetic hardware resources also require a complete mastery of dynamic computations: overflow may be signalled or truncation has to be performed. An efficient way to handle these problems is to use a direct binary representation. For example, this allows direct testing of the bit sign, performing of truncation or rounding, or displaying messages for under- or overflow.

To offer the same functionalities as for integers, arithmetic standard packages (cf. paragraph 6.3.2) provide overloaded arithmetic operations on vectors of the package STD_LOGIC_1164. Since it is necessary to specify the characteristic signed or unsigned of these vectors, two different types (SIGNED and UNSIGNED) are exported by the arithmetic package. Furthermore, a convention for the location of the most significant bit as well as the sign bit is defined (see paragraph 6.3.2.2). Arithmetic packages also usually propose conversion functions from and to integers. Indeed, a good arithmetic package is a powerful tool for modeling.

The following example illustrates such a use by modeling an adder (with input and output carries) on unsigned numbers:

```vhdl
library IEEE;
use IEEE.STD_LOGIC_1164.all;
use IEEE.NUMERIC_STD.all;
entity ADDER is
        generic (N : POSITIVE);
        port (CI        : in  STD_ULOGIC := '0';
              A, B      : in  UNSIGNED(N-1 downto 0);   -- Types exported from IEEE standard
              S         : out UNSIGNED(N-1 downto 0);   -- arithmetic package NUMERIC_STD
              CO        : out STD_ULOGIC                );
end ADDER;

architecture ARCH of ADDER is
begin
        P_ADD : process (CI, A, B)
            variable SUM : UNSIGNED(N downto 0);
            variable CARRY_IN : UNSIGNED (1 to 1);
        begin
            CARRY_IN (1) := CI;
            SUM := A + B + CARRY_IN; -- Two Overloaded "+" operators :
                              -- (UNSIGNED + UNSIGNED) and (UNSIGNED + STD_ULOGIC)
            CO  <= SUM(N);
            S   <= SUM(N-1 downto 0);
        end process P_ADD;
end ARCH;
```

The next example presents a subtraction of two unsigned numbers and performs a saturation to zero: the result is positive or zero.

```vhdl
entity SUBSTRACTION is
        generic (N : POSITIVE);
        port (A, B     : in  NATURAL range 0 to 2**N - 1;
              S        : out NATURAL range 0 to 2**N - 1   );
end SUBSTRACTION;

architecture ARCH of SUBSTRACTION is
        signal SUM : INTEGER range -2**N to 2**N - 1;
begin
        SUM <= A - B ;
        S <= 0   when SUM < 0 else       -- if the result is negative then it is saturated to zero
             SUM;                        -- else, it is considered as correct
end ARCH;
```

When considering the previous two examples, two distinct strategies emerge: the first one uses binary representation (bit vectors) whereas the second one deals with integers. What are the criterias for choosing between these two possibilities?

It seems obvious that the integer representation is generally more concise. Indeed, each time only a part of the object has to be considered (truncation, rounding, or saturation), a binary representation is more convenient. A good tradeoff may be to use the integer representation when it is possible, and to transform to binary representation using conversion functions exported by arithmetic packages when necessary.

Below are two examples performing truncation and rounding (only the N most significant bits are kept) on a sum of integers. Each of these examples uses a different internal representation: binary or integer based.

```vhdl
package P is
        constant P : POSITIVE := 12;
        subtype TYPE_I_P is INTEGER range -2**P to 2**P - 1;
        type TYPE_ARRAY_I is array (NATURAL range <>) of TYPE_I_P;
        constant N : POSITIVE := 8;
        subtype TYPE_I_N is INTEGER range -2**N to 2**N - 1;
end P;

use WORK.P.all;
entity ADDER is
        port (A      : in  TYPE_ARRAY_I ;        -- Input array, the integers to sum
              S_T    : out TYPE_I_N;             -- Truncated output
              S_R    : out TYPE_I_N          );  -- Rounded output
end ADDER;

-- Architecture using binary representation
library IEEE;
use IEEE.NUMERIC_BIT.all;    -- reference to the standard arithmetic package
architecture ARCH1 of ADDER is
begin
        P_ADD : process (A)
            variable SUM : TYPE_I_P;
            variable S     : SIGNED(P-1 downto 0);
            -- R is the half of the less significant bits (1/2 LSB) of the result
            constant R    : SIGNED(P-1 downto 0) := (P-N-1 => '1', others => '0');
        begin
            -- Computing the sum
            SUM := 0;
            for I in A'range loop
                SUM := SUM + A(I);
            end loop;
            -- Conversion into binary representation
            S := TO_SIGNED(SUM, S'length);
            -- Truncation
            S_T <= TO_INTEGER (S(P-1 downto P-N));
            -- Rounding
            if S(S'left) = '1' then     -- if S < 0
```

```
                    S := S - R;
            else
                    S := S + R;
            end if;
            S_R <= TO_INTEGER(S(P-1 downto P-N));
        end process P_ADD;
end ARCH1;


-- Architecture keeping the integer representation all along
architecture ARCH2 of ADDER is
begin
        P_ADD : process (A)
            variable SUM : TYPE_I_P;
            constant R    : TYPE_I_P := 2**(P-N-1);    -- 1/2 LSB (i.e. 2**(P-N)/2) of the result
        begin
            -- Computing the sum
            SUM := 0;
            for I in A'range loop
                SUM := SUM + A(I);
            end loop;
            -- Truncation
            S_T <= SUM / 2**(P-N);
            -- Rounding
            if SUM < 0 then
                SUM := SUM - R;
            else
                SUM := SUM + R;
            end if;
            S_R <= SUM / 2**(P-N);
        end process P_ADD;
end ARCH2;
```

Therefore, additions and subtractions have been seen in their general form. Indeed, a last specific form has to be studied: incrementation and decrementation. Their function is to add or respectively to subtract a unit value, and they are commonly associated with flip-flop memorization points. The unit value may be one, but also 2**N. This second possibility does not complicate the function too much: it only involves a scaling operation.

The first example presented below is a new version of a counter of the number of '1's in a bit vector already seen in a previous section (end of paragraph 4.1.2). The treatment is performed via a function that is general enough to be used in various contexts.

```
...
function NB1 (BV : in BIT_VECTOR) return NATURAL is
        variable V : NATURAL range 0 to BV'length;
begin
        V := 0;
        for I in BV'range loop
```

```
              if BV(I) = '1' then
                    V := V + 1;
              end if;
          end loop;
          return V;
   end NB1;
   ...

   -- Signal E is supposed to be an input signal of type BIT_VECTOR with a length of N
   -- Signal S is supposed to be an output signal of type NATURAL with the range 0 to N
   S <= NB1 (E);
```

The efficiency of such a description is not obvious. Taking the strict definition, N-1 incrementers are necessary if N bits are passed on as a parameter to function NB1. The synthesis tool will often perform Boolean optimizations in that case.

The last example of this section uses a process statement to describe a functionality very close to the previous one: counting the number of '0's or '1's of a given bit vector. Indeed, it is perfectly possible to write a function or a process for this purpose, as the choice of a process statement only aims to vary the modeling style.

```
entity COUNT is
      generic (B : BIT);   -- This parameter gives the value ('0' or '1') to be counted
      port (    BV : in  BIT_VECTOR;
                S   : out NATURAL        );
end COUNT;

architecture A of COUNT is
begin
      P_SEARCH : process
           variable VS : NATURAL range 0 to BV'length;
      begin
           VS := 0;
           for I in BV'range loop
                exit when BV(I) = not B;
                VS := VS + 1;
           end loop;
           S <= VS;
           wait on BV;
      end process P_SEARCH;
end A;
```

In this example, the type of a port is an unconstrained array. Indeed, the constraints of this vector will be decided when instantiating the model (direct instantiation of *VHDL'92*) or the corresponding component. In the same spirit, no constraint has been given to the output port S, which is of type NATURAL.

Connected to a signal of type NATURAL, the number of bits used for coding is that known by default for type NATURAL: usually 32 bits.

The declaration of the generic B of type BIT is not widely accepted by synthesis tools. This is mainly a question of maturity of these tools. A workaround consists in declaring a constant B of type BIT in a separate package declaration and to "use" (**use** clause) it. The drawback is that the value of this constant has to be declared within a package.

Additions and subtractions have been detailed in previous sections. Multiplications, divisions, and exponentiations have now to be discussed.

A large majority of synthesis tools propose multiplication as a current operation. Indeed, performances of these tools when synthesizing such operations may be very different. Some of them use very specific technologies (Wallace tree, carry save adder). Nevertheless, these operators are usually very expensive in terms of basic hardware resources, and thus in area. If a multiplication has to be fast with large size data (more than 8 bits), then it is not reasonable to perform it when targeting programmable devices (FPGA, EPLD). On the other hand, if the speed constraints are low, a solution that sequentializes the treatment in order to reuse a single adder has to be found. Synthesis tools are smart enough to optimize the required hardware resources in some specific cases, such as multiplication by a power of two.

The following example deals with multiplication between operands of different size. The result is truncated to a given size WZ. The assert statement aims to ensure the consistency between the different generics: size of operands and size of the result.

```
entity MULT is
        generic ( WX  : POSITIVE;        -- Size of operand X
                  WY  : POSITIVE;        -- Size of operand Y
                  WZ  : POSITIVE );      -- Size of result Z
        port (    X    : in  BIT_VECTOR(WX-1 downto 0);
                  Y    : in  BIT_VECTOR(WY-1 downto 0);
                  Z    : out BIT_VECTOR(WZ-1 downto 0)  );
begin
        assert WZ <= WX + WY report "Inconsistent Size" severity ERROR;
end MULT;

library IEEE;
use IEEE.NUMERIC_BIT.all;
architecture A of MULT is
begin
        Z <= BIT_VECTOR (
                        TO_UNSIGNED (
                            TO_INTEGER (UNSIGNED (X) * UNSIGNED (Y))
                            / 2**(WX + WY - WZ),
                        WZ)); -- Multiplication (not existing for BIT_VECTOR) and truncation
        end A;
```

Division is not supported by synthesis tools, except in an impaired form. Indeed, the reason for this restriction is simply the fact that the hardware resource implied by the general form of the division is not realistic. The impaired form universally supported is the division by a power of two.

The operations **rem** (remainder), **mod** (modulo), and the exponentiation (**) are also only supported if the right operand is a power of two.

### 4.1.4. Shift and Rotate Operations

These operations are usually performed on objects having a bit-vector representation. Two kinds of shift and rotate operations may be distinguished: logical and arithmetic (cf. paragraph 3.3.4.).

Arithmetic shifts have already been seen in the previous section. They may be used for multiplication (shift to the left) or division (shift to the right) if the right operand of such an operation is a power of two.

The following example illustrates a programmable shift right arithmetic. N is supposed to be a signal of type NATURAL with the range 0 to 3 and I and RESULT are signals of type INTEGER with the range -8 to 7.

```
RESULT <= I / 2**N;
```

The same behavior may be obtained using a bit vector representation instead of integers:

```
library IEEE;
use IEEE.STD_LOGIC_1164.all;
use IEEE.NUMERIC_STD.all;   -- Reference to the standard arithmetic package
...
-- N is a signal of type UNSIGNED with a length of 2.
-- I and RESULT are signals of type SIGNED with a length of 4.
RESULT <= I sra TO_INTEGER(N);
```

In this description, the operator **sra** (Shift Right Arithmetic, cf. paragraph 3.3.4.) is overloaded for left operand of type SIGNED. This predefined operator exists in *VHDL'92* only.

The conversion function TO_INTEGER transforms a type UNSIGNED into type INTEGER. This has been proposed for a long time by the proprietary package of synthesis tool vendors and is now provided by the standard arithmetic package (cf. paragraph 6.3.2.).

As shown in figure 4.3, the shift right arithmetic propagates the sign of the operand.

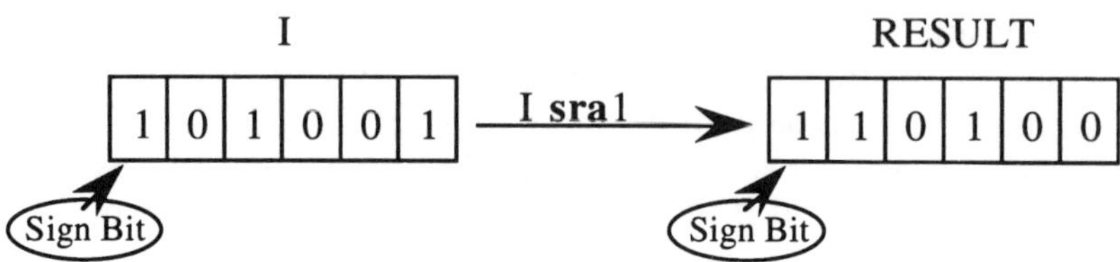


**Figure 4.3. (Single) Shift Right Arithmetic Operation**

VHDL also offers the possibility of describing the desired function more accurately. Using the slice-array signal assignment, all possible shifts may be described and therefore synthesized.

Shift right arithmetic is also described by the following lines:

```
library IEEE;
use IEEE.STD_LOGIC_1164.all;
use IEEE.NUMERIC_STD.all;    -- Reference to the standard arithmetic package
...

...
-- Signal N is of type NATURAL with a range 0 to 3.
-- Signals I and RESULT are of type SIGNED with a range 3 downto 0.
process (I, N)
        variable V : SIGNED(I'left -1 downto 0);
        alias SIGNE : STD_ULOGIC is I(I'left);
begin
        V := (others => SIGNE);
        RESULT <= V(N-1 downto 0) & I(I'left downto N);
end process;
...
```

Logical shifts include simple shifts as well as rotate operations. Both directions, left or right, are possible. To express them, two styles are usually available:

- The use of an assignment of bit-array slices.
- The use of the predefined shift operators (cf. paragraph 3.3.4.). These operators are predefined in *VHDL'92* and therefore directly usable for type BIT_VECTOR and are (will probably be) overloaded on type STD_LOGIC_VECTOR and STD_ULOGIC_VECTOR in a future version of package STD_LOGIC_1164.

The procedure below follows the first approach for describing a shift logical right operation:

```
procedure SHIFT_LOGICAL_LEFT (BV : inout BIT_VECTOR; N : in NATURAL) is

    -- The following line enforces the length, direction and index of BV bit representation
    alias A_BV : BIT_VECTOR(BV'length-1 downto 0) is BV;
    constant C : BIT_VECTOR(N-1 downto 0) := (others => '0');

begin
    assert N <= BV'length  report "N has to be less or equal to the length of BV"
                                severity ERROR;
    BV := A_BV(A_BV'left-N downto 0) & C;
end SHIFT_LOGICAL_LEFT;
```

The corresponding procedure call may only occur from within a sequential context (another subprogram or a process) due to the class of BV. BV is expected to be of class **variable**, default class for parameters of mode **out** or **inout**. Thus, this procedure only handles *variables* of type BIT_VECTOR.

If designers want to call it from a concurrent context, two changes have to be done in the procedure code. Parameter BV has to be explicitly specified of class **signal** and the assignment symbol within the procedure body has to be changed from ":=" to "<=". The procedure declaration becomes

```
procedure SHIFT_LEFT (signal BV : inout BIT_VECTOR; N : in NATURAL);
```

Of course, it is highly recommended to use the predefined *VHDL'92* shift operators or equivalent functions exported by synthesis specific packages whenever possible. In that case, descriptions become short and sure.

A shift right logical operation is thus:

```
...

RESULT <= I srl N;              -- With VHDL'92
...

-- With VHDL'87
library IEEE;
use IEEE.???.all;               -- Reference to a package specific to the synthesis tool
...
RESULT <= SRL (I, N) ;          -- SRL is a function of the previous package.
...
```

It may unfortunately happen that the data type is not a bit vector. Local conversions are therefore necessary to return to the previous cases. If the object to be shifted is positive or equal to zero, a multiplication or division by a power of two may be envisaged. In that case, there is no difference between a logical and an arithmetic shift. The following example illustrates a right logical shift on an object of type INTEGER.

```
-- I is a variable of type INTEGER,
-- RESULT is a signal of type INTEGER
-- N is a constant of type NATURAL with the value 2
...
library IEEE;
use IEEE.NUMERIC_BIT.all; -- Access to functions TO_INTEGER and TO_UNSIGNED
-- The context is mandatory sequential (process or subprogram)
 -- VHDL'92 operator is used
RESULT <= TO_INTEGER(TO_UNSIGNED(I, RESULT'length) srl N);
```

Indeed, three different functions or operator calls are embedded in this single sequential signal assignment:

- The conversion function TO_UNSIGNED is first executed and performs the translation of the integer I into an object of type UNSIGNED with the same length as RESULT.
- Then, the shift operator **srl** performs the desired operation on the previous object of type UNSIGNED.
- Finally, the conversion function TO_INTEGER transforms the object of type UNSIGNED into an integer compatible with the type of RESULT.

## 4.1.5. Multiplexers

In electronic system architectures, the hardware devices handling the connections between the hardware resources are critical. In the state of the art of today's synthesis tools, only a very small set of such devices is available. In the near future, architectures based on busses should be available.

Multiplexing and demultiplexing are among the functions supported by today's synthesis tools. The basic functionality of a multiplexer is to select one input among several whereas a demultiplexer performs the opposite operation: one input is propagated to one of the outputs and the other outputs keep their previous values. Therefore, a demultiplexer usually implies a memorization. Since this section is devoted to purely combinational hardware resources, demultiplexers, despite their similitude with multiplexers will be studied later with other synchronous hardware elements.

This first example shows the basic function of a multiplexer: one of the inputs is propagated to the single output. A logic equation can describe such a functionality:

```
entity MUX2_TO_1 is
        port (    A, B          : in  BIT;
                  SELECT_A  : in  BIT;
                  Z             : out BIT);
end MUX2_TO_1;
```

```
architecture FIRST of MUX2_TO_1 is
begin
        Z <= (A and SELECT_A) or (B and not SELECT_A);
end FIRST;
```

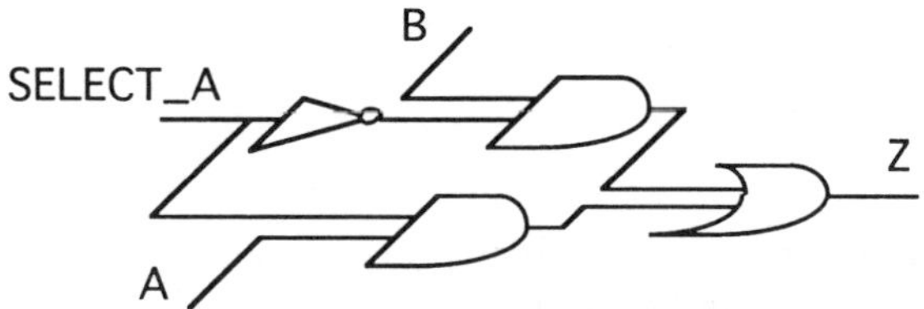


**Figure  4.4. Synthesized  Result  of  Combinatorial  Multiplexer**

Another  general  method  for  describing  multiplexers  consists  in
conditionally  to  the  command  assigning  the  inputs  to  the  single  output. The
following three styles (architectures SECOND, THIRD and FOURTH) clearly
express the same functionality:

```
-- Using the concurrent signal assignment in its conditional form
architecture SECOND of MUX2_TO_1 is
begin
        Z   <=   A    when SA = '1' else
                 B;
end SECOND;
```

**Figure  4.5. Usual  Representation  of  a  Multiplexer**

```
-- Using the concurrent signal assignment in its selected form
architecture THIRD of MUX2_TO_1 is
begin
        with SELECT_A select
            Z   <=   A when '1',
                     B when '0';
end THIRD;


-- Using a sequential signal assignment. This is the equivalent process of architecture SECOND.
architecture FOURTH of MUX2_TO_1 is
begin
```

```
      P_MUX : process (A, B, SELECT_A)
      begin
          if SELECT_A = '1' then
              Z <= A;
          else
              Z <= B;
          end if;
      end process P_MUX;
  end FOURTH;
```

Indeed, the description of multiplexers may be done using many other styles based on the following two ideas:

- If the data to be selected are not structured (and the previous four examples are in this category), a selector determines the data to be assigned to the output. The type of selector has to be enumeratable and then codable. Therefore, enumerated types (BIT, STD_LOGIC, etc.) but also type INTEGER and its associated subtypes, as well as composite types (BIT_VECTOR, etc.) of these types are accepted.

  Input and output types have to be compatible. Indeed, it should be possible to directly assign the inputs to the outputs. The statements used for the description of such hardware resources are the conditional or selected forms of the concurrent signal assignment or, when using an explicit process, the sequential **if** and **case** statement. In this latter case, the output has to be the target of an assignment within each branch of the statement.

  The main interest of this style of description is the possible use of complex functions in the assignment. As shown in paragraph 4.1.6. ALU, important operative hardware resources may possibly be inferred by such a statement.

- If the data to be selected are structured, a composite type should be defined to represent the entire input data. The types of elements of this composite type have to be compatible with each other and, of course, with the output type. The type of index, or indices, of this data structure has itself to be enumeratable, just like the selector used in the previous description style.

  Concision and readability are the main advantages of this description style: the entire description results in a single assignment. Indeed, the data structure is necessarily computed before and the concision is not as efficient as it appears at first glance. For example, it is not obvious to mix multiplexing and operations that do not assign input data as a whole.

Here is an example to illustrate the two styles discussed:

```
-- Given the type TYPE_TEMPERATURE defined as follows:
type TYPE_TEMPERATURE is (VERY_COLD, COLD, WARM, HOT, VERY_HOT);
```

The first modeling style is:

```
-- O, I1, I2, I3 are supposed to have the same type and
-- TEMPERATURE is a signal of type TEMPERATURE
process (TEMPERATURE, I1, I2, I3)
begin
        case TEMPERATURE is
            when VERY_COLD     =>  O <= I1;
            when COLD          =>  O <= I1;
            when WARM          =>  O <= I2;
            when HOT           =>  O <= I3;
            when VERY_HOT      =>  O <= I3;
        end case;
end process;
```

Putting together the common treatments, another form provides a more concise code.

```
process (TEMPERATURE, I1, I2, I3)
begin
        case TEMPERATURE is
            when VERY_COLD I COLD     =>  O <= I1;
            when WARM                 =>  O <= I2;
            when HOT I VERY_HOT       =>  O <= I3;
        end case;
end process;
```

Many other forms are possible. The most important thing is that signal O is assigned whatever the value of the selector: TEMPERATURE.

Using the second modeling style, a data structure allowing indexing is required. Here are the details:

```
-- TYPE_I is the type of the data. Its accurate definition is not important in this example.
-- TYPE_ARRAY_I is the data structure indexed with TYPE_TEMPERATURE
type TYPE_ARRAY_I is array (TYPE_TEMPERATURE) of TYPE_I;
```

If we suppose ARRAY_I is a signal of type TYPE_ARRAY_I, the description in the concurrent domain is

```
ARRAY_I  <= (I1, I1, I2, I3, I3);
O        <= ARRAY_I (TEMPERATURE);
```

These two ways of describing a multiplexer are illustrated in a second example: a multiplex tree. A multiplex tree is a structured and hierarchical description using elementary multiplex functions.

Below is the first style, without structured data:

```vhdl
package P is
        -- Declaration of the data subtype: It guarantees the consistency between input and output
        -- types. The alternative is to declare a generic parameter.
        subtype BV4 is BIT_VECTOR(3 downto 0);
        -- Declaration of an enumerated type for the selector.
        type TYPE_PRIORITY is (LOW, MEDIUM, HIGH);
end P;

use WORK.P.all;
entity MUX_TREE is
        port (  IN0_LOW, IN0_MEDIUM, IN0_HIGH    : in  BV4;
                IN1_LOW, IN1_MEDIUM, IN1_HIGH    : in  BV4;
                IN2_LOW, IN2_MEDIUM, IN2_HIGH    : in  BV4;
                LINE_NUMBER                      : in  NATURAL range 0 to 2;
                PRIORITY                         : in  TYPE_PRIORITY;
                OUT0                             : out BV4               );
end MUX_TREE;
```

First architecture using concurrent assignments:

```vhdl
architecture A of MUX_TREE is
        signal S_MEDIUM, S_LOW, S_HIGH  : BIT_VECTOR(OUT0'range);
begin
        -- First level of selection
        with LINE_NUMBER select
                S_LOW <=        IN0_LOW         when 0,
                                IN1_LOW         when 1,
                                IN2_LOW         when 2;
        with LINE_NUMBER select
                S_MEDIUM <=     IN0_MEDIUM when 0,
                                IN1_MEDIUM when 1,
                                IN2_MEDIUM when 2;
        with LINE_NUMBER select
                S_HIGH <=       IN0_HIGH        when 0,
                                IN1_HIGH        when 1,
                                IN2_HIGH        when 2;

        -- Second level of selection
        with PRIORITY select
                OUT0 <=         S_LOW           when LOW,
                                S_MEDIUM        when MEDIUM,
                                S_HIGH          when HIGH;
end A;
```

Of course, the same functionality may be described using (sequential) case statements:

```vhdl
architecture B of MUX_TREE is
begin
     P : process ( LINE_NUMBER, IN0_LOW, IN0_MEDIUM, IN0_HIGH, IN1_LOW,
                   IN1_MEDIUM, IN1_HIGH, IN2_LOW, IN2_MEDIUM, IN2_HIGH)
     begin
     -- Second level of selection
        case LINE_NUMBER is
           when 0  =>  case PRIORITY is        -- First level of selection
                          when LOW         =>  OUT0 <= IN0_LOW;
                          when MEDIUM      =>  OUT0 <= IN0_MEDIUM;
                          when HIGH        =>  OUT0 <= IN0_HIGH;
                       end case;
           when 1  =>  case PRIORITY is        -- First level of selection
                          when LOW         =>  OUT0 <= IN1_LOW;
                          when MEDIUM      =>  OUT0 <= IN1_MEDIUM;
                          when HIGH        =>  OUT0 <= IN1_HIGH;
                       end case;
           when 2  =>  case PRIORITY is        -- First level of selection
                          when LOW         =>  OUT0 <= IN2_LOW;
                          when MEDIUM      =>  OUT0 <= IN2_MEDIUM;
                          when HIGH        =>  OUT0 <= IN2_HIGH;
                       end case;
        end case;
     end process P;
end B;
```

This last description has a real advantage on the previous concurrent one: it allows for nesting case statements with any number of levels without using, and therefore defining, intermediate objects (signals S_LOW, S_MEDIUM, S_HIGH in the first example).

Structuring the data to make the selection easier, the same functionality becomes:

```vhdl
package P is

        -- Declaration of the data subtype.
        subtype BV4 is BIT_VECTOR(3 downto 0);
        -- Declaration of an enumerated type for the selector.
        type TYPE_PRIORITY is (LOW, MEDIUM, HIGH);
        -- Declaration of the structured data.
        type TYPE_ARRAY_P is array (TYPE_PRIORITY) of BV4;
        type TYPE_ARRAY_IN is array (NATURAL range <>) of TYPE_ARRAY_P;

end P;
```

```
use WORK.P.all;
entity MUX_TREE is
        port (    IN0                 : in  TYPE_ARRAY_IN;
                  LINE_NUMBER    : in  NATURAL range 0 to 2;
                  PRIORITY          : in  TYPE_PRIORITY;
                  OUT0               : out BV4                    );
end MUX_TREE;

architecture A of MUX_TREE is
begin
        OUT0 <= IN0 (LINE_NUMBER) (PRIORITY);
end A;
```

Indeed, the complexity of the description is now in the data type declarations. However, the inferred hardware will be the same for all the proposed modeling styles.

In certain cases, it is necessary to keep more flexibility: the dimension of the data has to be parametrized. The last studied description does not allow for such liberty because an array element has to be constrained. In case of changes in the size of data, package P has to be modified. Furthermore, two instantiations of this model with different sizes are not possible or imply very significant changes such as duplications of the descriptions with all the potential conflicts of names.

The only realistic solution is the use of a generic parameter providing the size of the input and output data. The data structure definition must be such that the base element is constrained. Since type BIT_VECTOR is unconstrained and therefore type BIT is the best candidate. This implies to handle a two-dimension array of bits instead of a one-dimension array of bit vectors. The indices of the structure are
   • the index for line selection (LINE_NUMBER)
   • the index for priority selection (PRIORITY)
   • the index for the position within the word.
This index varies from 0 to N-1, where N is the generic parameter providing the size of data.

The description becomes

```
package P is
        -- Declaration of an enumerated type for the selection.
        type TYPE_PRIORITY is (LOW, MEDIUM, HIGH);
        -- Declaration of the structured data.
        type TYPE_ARRAY_IN is array (    NATURAL range <>,          -- LINE_NUMBER
                                         TYPE_PRIORITY range <>,    -- PRIORITY
                                         NATURAL range <>           -- BIT_VECTOR
                                    ) of BIT;
end P;
```

```vhdl
use WORK.P.all;
entity MUX_TREE is
        generic ( N : POSITIVE;      -- Size of (output) datas
                  L : POSITIVE );    -- Maximal number of lines
        port (    IN0               : in  TYPE_ARRAY_IN(   0 to L-1,
                                                           TYPE_PRIORITY,
                                                           N-1 downto 0          );

                  LINE_NUMBER       : in  NATURAL range 0 to L-1;
                  PRIORITY          : in  TYPE_PRIORITY;
                  OUT0              : out BIT_VECTOR(N-1 downto 0)    );
end MUX_TREE;

architecture A of MUX_TREE is
begin

        process (IN0, LINE_NUMBER, PRIORITY)
        begin
            for I in OUT0'range loop
                OUT0(I) <= IN0 (LINE_NUMBER, PRIORITY, I);
            end loop;
        end process;

end A;
```

The designer has to be aware that this last style is not always a possibility. The use of a specific data structure (last description) is only possible when the selection function does not imply specific treatment on such or such element. If **case** or **select** can be used, then this approach is possible.

In the other cases, conditional statements have to be used. In the following example, the use of a conditional statement is strongly recommended:

```vhdl
-- I0, I1, I2, I3 and O are signals with the same type and
-- INIT, COND1, COND2 et COND3 are signals of type BOOLEAN.
process
begin
        if INIT then
            O <= I0;
        elsif COND1 or COND2 then
            O <= I1;
        elsif not COND3 then
            O <= I2;
        else
            O <= I3;
        end if;
        wait on INIT, COND1, COND2, COND3, I0, I1, I2, I3;
end process;
```

126 Circuit Synthesis with VHDL

Although the **if** statement is recommended in the previous case, it is always possible to use a case statement instead. In that case, the only issue is to code the information in a bit-level representation, possibly using package ST_LOGIC_1164, which provides a "don't care" value ('-').

The following example describes the same functionality as previously. Indeed, the only difference consists in changing the type of selection. Type STD_LOGIC has been chosen mainly because it fits as well with conditional as with selected statements: a bit vector may be used for selection. First is given the conditional form followed by the selected one:

```
-- Conditional form:
-- I0, I1, I2, I3 and O are signals of the same type and
-- INIT, COND1, COND2 and COND3 are signals of type STD_ULOGIC.
O <= I0    when INIT= '1'                      else
     I1    when COND1='1' or COND2='1'  else
     I2    when COND3='0'                    else
     I3;


-- Selected form:
-- Subtype BV4 is declared as followed to allow the recognition of the four commands as
-- bit vectors.
subtype BV4 is STD_ULOGIC_VECTOR(0 to 3);
-- ...
-- Somewhere, in a concurrent context:
with BV4'(INIT & COND1 & COND2 & COND3) select
     O <=    I2    when    "0000",
             I1    when    "011-",
             I0    when    "1---",
             I3    when    others;
```

These two examples illustrate a very important point: conditional and selective forms are not really equivalent, and the result (after synthesis) may be different. Conceptually, a conditional statement implies a priority in the treatment. In a selective form, accesses to branches have the same priority.

Therefore, in the conditional form, the condition INIT='1' is first evaluated; then, if INIT='0', the condition COND1='1' **or** COND2 = '1' is evaluated, etc. The **else** clause is the negation of all the Boolean expressions previously expressed. Therefore, the inferred hardware is not a simple multiplexer but much more a priority encoder.

When the designer can choose between conditional and selective forms, the latter has often to be chosen: conditional forms may significantly decrease the speed performance. As a matter of fact, conditions in a conditional form imply more logic levels to be decoded. The same description using selective statements would inferred a more balanced decoding.

**Remark:**
"Don't care" values ('-') are used in the previous example. When comparing a "don't care" value to any different other value, the simulation concludes that they are different. This is not what is expected from the synthesis semantics of the "-" character and may imply differences between the simulation of the description before synthesis and the simulation of the synthesis result. This point is discussed in section 5.1.

The next example illustrates this problem of multiplexing and more precisely the use of selective versus conditional forms. No notion of priority is supposed to exist in the following case. Here is the conditional form:

```
-- I1, I2, I3, I4 et O are signals with the same data type and
-- SEL is a signal of type BIT_VECTOR with a length of 2.
O <=  I1 when  SEL = "00"   else
      I2 when  SEL = "01"   else
      I3 when  SEL = "10"   else
      I4;
```

The selective form follows:

```
with SEL select
O <=  I1 when  "00",
      I2 when  "01",
      I3 when  "10",
      I4 when  "11";
```

## 4.1.6. ALU

Arithmetic and logical units are commonly used in microprogrammable architectures. Each time several logical or arithmetic operations are necessary but can be accessed at different (cycle) times, these resources are of real interest: they allow a significant gain in hardware area. Nevertheless, the designer has to manage in time the sequencing of the control of these resources.

In chapter 3, the resource sharing issue is overviewed. An arithmetic resource (an adder for example) may be shared provided its use is exclusive in the time. A classical example of this situation is

```
-- A, B, C, D, O are signals with the same type and CMD is a signal of type BOOLEAN.
-- The following description is supposed to be within a concurrent context (architecture or block).
O <=  A + B    when CMD else
      C + D;
```

The following choice is open:
- Using two adders and a multiplexer (two to one) for the selection of the result
- Using two multiplexers on the inputs for selecting the right pair (A, B or C, D) and a single adder

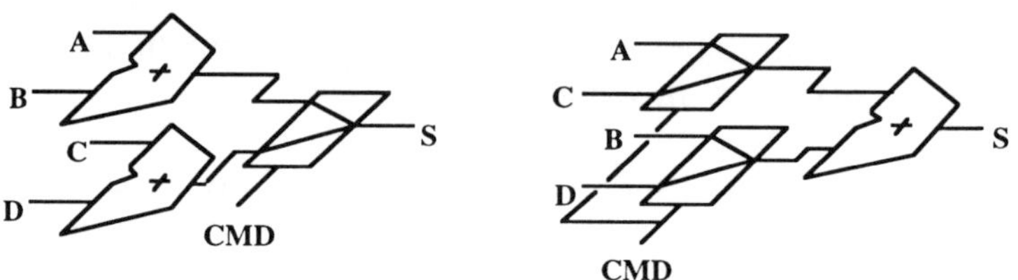


**Figure 4.6. Resource Sharing, Adder**

The latter solution obviously involves lower area consumption.

As shown by the next two examples, resource sharing may also be performed when operations are different but belong to the same "family." In that case, simple ALU may be involved:

Example 1:

```
-- A, B, O are signals with the same type (INTEGER range ...).
-- The following description is supposed to be within a concurrent context (architecture or block).
S <=   A + B when CMD else
       A - B;
```

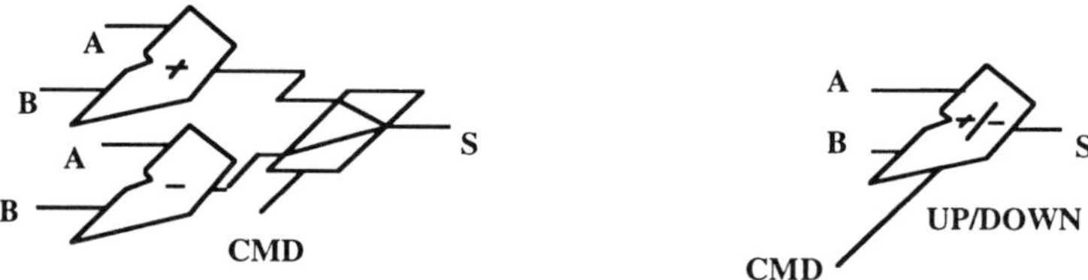


**Figure 4.7. Resource Sharing 1, Simple ALU**

Example 2:

```
-- A, B, C, D, S are signals with the same type (INTEGER range ...). CMD is a boolean signal.
-- The following description is supposed to be within a concurrent context (architecture or block).
S <=   A + B when CMD else
       C - D;
```

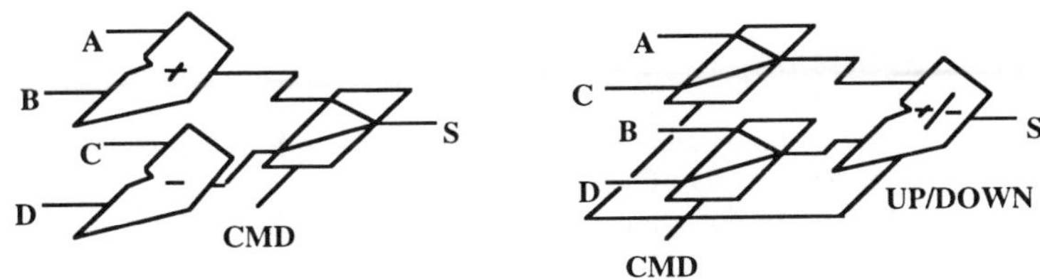


**Figure 4.8. Resource Sharing 2, Simple ALU**

Here the need for an adder and a subtractor is clear, but its use is exclusive in time. Therefore, the synthesis tool is allowed to infer a single adder-subtractor operator. This simple example shows that synthesis tools are able to infer simple ALUs in obvious cases. Nevertheless, in the general case, the designer has to explicitly describe the set of resources he or she wants to be shared. The next example illustrates this:

```vhdl
package P is
        type TYPE_OP is (ADD, SUB, INC0, DEC0, ABS0, ABS1);
end P;

use WORK.P.all;
library IEEE;
use IEEE.STD_LOGIC_1164.all;
use IEEE.NUMERIC_STD.all;
entity ALU is
  generic (N : POSITIVE);
  port (    OP       : in  TYPE_OP;
            I0, I1   : in  SIGNED(N-1 downto 0);
            O        : out SIGNED(N-1 downto 0)      );
end ALU;

architecture A of ALU is
begin
        P_ALU : process
            -- Function declaration, local to the process, which computes the absolute value.
            function F_ABS (  signal I : in SIGNED) return SIGNED is
                variable V : SIGNED(I'range); -- Declaration used to size the result
            begin
                if TO_INTEGER(I) >= 0 then
                    V := I;
                else
                    V := TO_SIGNED (0, V'length) - I;
                end if;
                return V;
            end F_ABS;
        begin
```

```
                case OP is
                    when ADD    =>   O <= I0 + I1;
                    when SUB    =>   O <= I0 - I1;
                    -- Use of the overloaded "+" operator between SIGNED and INTEGER types
                    when INC0   =>   O <= I0 + 1;
                    -- Use of the overloaded "-" operator between SIGNED and INTEGER types
                    when DEC0   =>   O <= I0 - 1;
                    when ABS0   =>   O <= F_ABS(I0);
                    when ABS1   =>   O <= F_ABS(I1);
                end case;
                wait on OP, I1, I0;
            end process P_ALU;
    end A;
```

The previous architecture is purely arithmetic: all operations are very close
to the addition. This kind of description allows the synthesis tool to achieve an
efficient optimization.

The interest of a separate definition of function F_ABS has to be noted: it
avoids any code duplication. Of course, this structuration does not modify the
final synthesis result.

The following example now explores shift and rotate operations:

```
package F_PKG is
        function F_ROUND(NAT : in NATURAL) return NATURAL;
end F_PKG;


library IEEE;
use IEEE.NUMERIC_BIT.all;
package body  F_PKG is
        function F_ROUND(NAT : in NATURAL) return NATURAL is
        begin
            if NAT mod 2 = 0 then
                return NAT/2;
            else
                return NAT/2 + 1;
            end if;
        end F_ROUND;
end F_PKG;


library UTIL_LIB;
use UTIL_LIB.F_PKG.all;
entity ALU is
  generic ( N : POSITIVE);
  port (    I    : in  BIT_VECTOR(N-1 downto 0);
            D    : in  NATURAL range 0 to F_ROUND(N/2);
            C    : in  BIT_VECTOR(0 to 1);
            O    : out BIT_VECTOR(N-1 downto 0)    );
end ALU;
```

```
architecture A of ALU is
begin
        with C select
            O <=      |    srl   D    when "00",
                      |    sll   D    when "01",
                      |    rrl   D    when "10",
                      |    rll   D    when "11";
end A;
```

Two points are illustrated in the above example:
- This description shows the use (clauses **library** and **use**) of a user-defined package stored in a library that is different from WORK: UTIL_LIB in this case.
- The previous source code also points out that constraint on objects, here on port D, may be computed. To be allowed to do so, the constraint has to be globally static. This means that its value is known once the netlist is built (i.e., at the end of the elaboration phase). In our case, N may be considered as locally constant: its value is given either when instantiating or when configuring. Therefore, the returned value of function F_ROUND is known at the same time. Unfortunately, many tools do not accept such a description, and the designer has to write a less general (generic) description. In that case, a generic parameter may be replaced by two other ones, charge to the designer to ensure an implicit consistency.

In our example, the entity declaration should provide two generic parameters instead of one. The first one gives the value of N and the second one F is (supposed) to be equal to F_ROUND(N/2). The consistency has to be ensured at a upper level, during configuration or instantiation. This may really lead to a trap: the two parameters seem independent from within the entity but they are not. As the following shows, a good practice is to add an appropriate comment and a concurrent assert statement to check the relationship between the two generic parameters. Of course, this statement has no synthesis semantics.

```
library UTIL_LIB;
use UTIL_LIB.F_PKG.all;
entity ALU is
    generic ( N : POSITIVE;
              F : POSITIVE);    -- F is equal to F_ROUND(N/2)
    port (   I    : in  BIT_VECTOR(N-1 downto 0);
             D    : in  NATURAL range 0 to F;
             O    : out BIT_VECTOR(N-1 downto 0)   );
begin
    assert F=F_ROUND(N/2) report "Bad Consistency Check between N and F";
end ALU;
```

### 4.1.7. Three-State Logic

Three-state buffers are necessary whenever several blocks concurrently access a single bus. For modeling such hardware resources, few possibilities are indeed available.

On the one hand, VHDL defines the notions of guarded block and resolved signal, which have been put into the language for this modeling purpose. On the other hand, the designers (represented by the IEEE) have standardized a package exporting a nine-state logic type and its relevant functions and operators. Among these nine values, 'Z' is proposed for bus mechanism modeling. The package STD_LOGIC_1164 defines the resolved type STD_LOGIC as well as type STD_LOGIC_VECTOR, vector of elements of type STD_LOGIC. The associated resolution function, also defined and exported by this package, gives to 'Z' the semantics of "three-state" value.

Therefore, two different styles are possible for modeling three-state logic. Here is described the use of package STD_LOGIC_1164, which is the simplest and the most popular to describe a bidirectional three-state resource. Indeed, the following source code is describing the resource illustrated by figure 4.9. Such a device may adapt any hardware with an input and an output to a three-state bus mechanism.

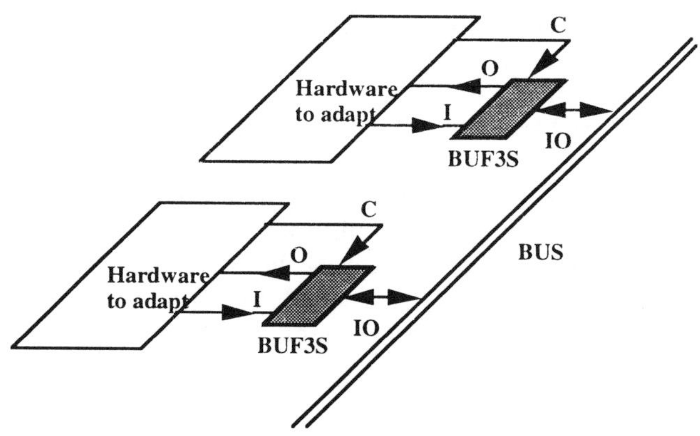


**Figure 4.9. Bidirectional Three-State Resource**

```
library IEEE;
use IEEE.STD_LOGIC_1164.all;
entity BUF3S is
        generic (N : POSITIVE);
        port (   I    : in      STD_ULOGIC_VECTOR(0 to N-1);
                 O    : out     STD_ULOGIC_VECTOR(0 to N-1);
                 C    : in      STD_ULOGIC;
                 IO   : inout   STD_LOGIC_VECTOR(0 to N-1)   );
end BUF3S;
```

```
architecture A of BUF3S is
begin
      IO   <=   STD_LOGIC_VECTOR(I) when      C = '1'
                   else (others => 'Z');
      O   <=   STD_ULOGIC_VECTOR(IO);
end A;
```

Using this first style, the designer has to declare a resolved signal and to assign the value 'Z' each time the output is no longer driven.

The second style is quite different:

```
library IEEE;
use IEEE.STD_LOGIC_1164.all;
entity BUF3S is
      generic (N : POSITIVE);
      port (   I    : in       STD_ULOGIC_VECTOR(0 to N-1);
               O    : out      STD_ULOGIC_VECTOR(0 to N-1);
               C    : in       STD_ULOGIC;
               IO   : inout     STD_LOGIC_VECTOR(0 to N-1) bus );
end BUF3S;

architecture A of BUF3S is
begin
      BG : block (C = '1')
      begin
            IO   <=   guarded STD_LOGIC_VECTOR(I);
      end block BG;
      O   <=   STD_ULOGIC_VECTOR(IO);
end A;
```

Here, the signal has not only to be of a resolved type but also to be guarded with the option **bus**. Therefore, each time all its sources are disconnected (all commands are inactive), the value on the bus IO will be those returned by the resolution function RESOLVED of package STD_LOGIC_1164 when called with a null input array. When all commands are inactive, the returned resolved value is 'Z'.

This last way of describing three-state logic, even if logically equivalent to the previous one, is generally not well accepted by synthesis tools. It is true that it significantly increases the complexity of the description and is therefore less obvious to use with success.

This concept of connection/disconnection, well defined within the language, is not really popular among designers. A lot of issues are still open: is it possible to use it for synthesis with a type other than STD_LOGIC (or related)? For example, will a resolved integer type be accepted? Nothing in theory prohibits such a description style for three-state logic hardware. The main problem comes from the choice of bus disconnection value. With a logic type, a particular value ('Z') is dedicated to this purpose. Using an integer type, the choice of disconnection value is difficult to specify. It may even be possible that no integer values are available for coding this state: all integer values have another consistent meaning. Do we have to choose the smallest value? zero? Indeed, no choice is perfect and the main drawback is that this value will always be different in simulation and after synthesis where everything is described at a bit-level representation.

## 4.1.8. ROM

Several different styles have been introduced for describing logic equations: simple assignments of logic expressions, conditional assignments, etc. These statements are usually synthesized using medium complexity gate netlists. When the hardware has to provide a more complex logic function, ROMs may be used.

ROMs may be considered as truth tables representing the behavior. The output values are defined for all possible input values. The interest of such a resource is to avoid the explicit description between inputs and outputs. Indeed, such a function may be too complex to be synthesized as a gate netlist. Since certain specifications may not be expressed as algorithms, a ROM description is really useful. One of the optimum domains for using a ROM is coding and decoding.

Since a ROM is only a table expressing outputs as a function of inputs, its useful information is generally redundant. Therefore, this kind of description is well suited to the knowhow of synthesis tools: one of their main functions is to reduce and minimize logic equations. Indeed, it is even perfectly possible that after optimization of a function that appears to be complex, the synthesis tool decides to synthesize it as a simple gate netlist instead of a real ROM.

Modeling a ROM consists in defining an array of values. The array index is the ROM input. Its type is necessarily an integer or an enumerated type. The array element is the ROM output. No specific constraint is required on that type and any type allowed in synthesis (see section 3.2.) may be used.

Here is a simple example of ROM modeling:

```vhdl
-- Type of the index (i.e. of the ROM input)
type TYPE_CMD is (LEFT, RIGHT, TOP, BOTTOM, MIDDLE);
-- The ROM input signal
signal CMD : TYPE_CMD;
-- Type of the ROM content
type TYPE_A_CMD is array (TYPE_CMD) of BIT_VECTOR(0 to 3);
constant CMD_FUNCTION : TYPE_A_CMD :=    -- ROM content
                    (   LEFT        => "1000",
                        RIGTH       => "0001",
                        TOP         => "0100",
                        BOTTOM      => "0010",
```

```
                       MIDDLE        => "1111"    );
-- Declaration of the output signal with a type consistent with the range of the array
-- elements of CMD_FUNCTION.
signal OUTPUT : BIT_VECTOR(CMD_FUNCTION(CMD_FUNCTION'left)'range);

-- Concurrent statement giving the relationship between ROM input (CMD) and its output
-- (OUTPUT).
OUTPUT <= CMD_FUNCTION (CMD);
```

The following example of ROM deals with integer types. Its purpose is to
store the results of a sine function of an angle value comprised between 0 and
$\pi/2$. The angle value, input of the ROM, is coded using P bits. The resulting
value is defined with a K bit width. This example illustrates the way to model
functions that are not straightforward to achieve in hardware.

```
library MATH_LIB;
use MATH_LIB.TRIGO_PKG.all;
-- Function SIN, as well as constant PI, are imported from this package

...
subtype NAT_P is NATURAL range 0 to 2**P-1;
subtype NAT_K is NATURAL range 0 to 2**K-1;

type TYPE_A_INTEGER is array (NATURAL range <>) of NAT_K;
function CONV_SIN (K, P : in NATURAL) return TYPE_A_INTEGER is
        -- Function SIN is defined as follows: function SIN(X : in REAL) return REAL;
        -- where R represents the angle expressed in radians. The returned value is a real
        -- between 0.0 and 1.0.
        variable R       : TYPE_A_INTEGER(0 to 2**P-1);
        variable V_REAL : REAL;
begin
        for I in R'range loop
            V_REAL := SIN ( REAL(I) / 2.0**P * PI / 2.0);
            R(I) := INTEGER ( V_REAL * 2.0**K );
        end loop;
        return R;
end CONV_SIN;

-- Declaration of the constant array storing the sine value.
constant SIN_TABLE : TYPE_A_INTEGER := CONV_SIN(K, P);
-- Declaration of input/output signals
signal INPUT        : NAT_P;
signal OUTPUT      : NAT_K;

...

-- Concurrent assignment expressing the relationship between inputs and outputs
OUTPUT <= SIN_TABLE (INPUT);
```

It is important to note that the constant SIN_TABLE is dynamically computed. It depends on the values of two parameters, K and P, which are static and known at compilation time. As a result, SIN_TABLE is a real constant throughout the execution. Nevertheless, this kind of description where a constant is computed during elaboration (SIN_TABLE) is rarely accepted by synthesis tools. The current tool generation usually only accepts constants whose values are known just after analysis, and thus before elaboration.

One way of working around this lack of maturity of synthesis tools is to generate the already computed value for each given pair (K, P). This operation may be performed manually if the number of pairs is small but should be automated in the opposite case. Indeed, a VHDL (or C, Ada, etc.) program may be written for this VHDL code generation.

## 4.1.9. PLA

PLAs, Programable Logic Arrays, are together with ROMs, a way of representing logic equations. The fundamental difference between ROMs and PLAs is in the organization and structure of the logic equations. Indeed, each output of a PLA may be expressed as the sum of input products. This (Boolean) sum is expressed using the **or** operator whereas the (Boolean) product involves the **and** operator. PLAs are usually characterized by three parameters:
  • the number of binary inputs (P)
  • the number of binary outputs (K)
  • the number of intermediate products, also called monoms (M)

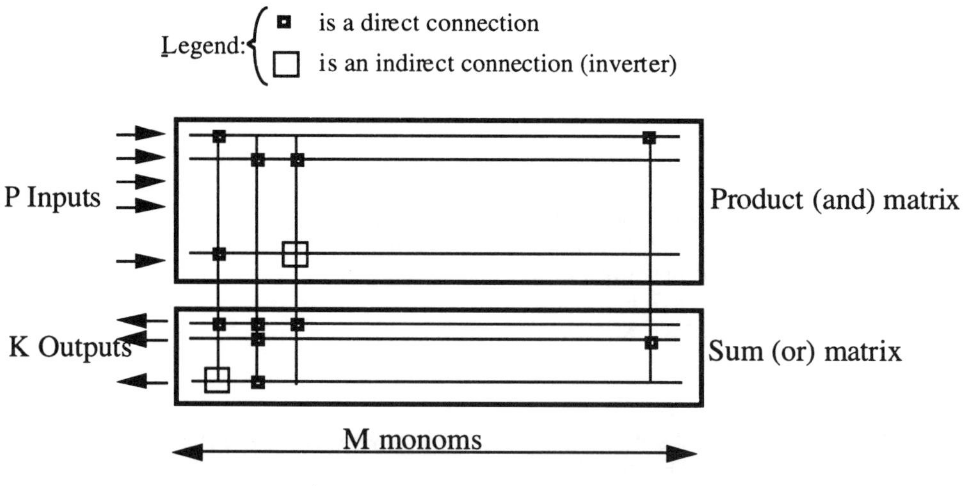


**Figure  4.10.        PLA  Principle**

A shown in figure 4.10, a PLA may be represented by two matrixes. Within the product matrix, a binary input may be used, directly or after having been inverted, for each monom. The same principle is used for the sum matrix which involves up to M intermediate or partial products.

Describing a function in the form of a PLA is no easy task in VHDL. Many other much more straightforward ways of expressing functions are possible. Nevertheless, being able to interpret a PLA description may be of interest to ensure the connection with other synthesis tools (for example based on automatons) or to restart from a description explicitly using PLAs.

In addition to illustrating a PLA modeling, the next example shows a general and abstract synthesis style that may be used in many other cases. In this example, the programming of the PLA is achieved, for mnemonic reasons, using only three values of the integer type: 0 when no connection, 1 when direct connection, and -1 when indirect connection.

```vhdl
-- Declaration of types describing the structure of the PLA array.
subtype TYPE_ELEMENT is INTEGER range -1 to 1;
type TYPE_MATRIX is array (NATURAL range <>, NATURAL range <>)
                of TYPE_ELEMENT;


-- Declaration of the function returning the M products:
--      I is the input.
--      MAT is the matrix of products. Its first index has to be consistent with the input range
--      and Its second index with the number of monoms.
function F_AND_PLA(I : BIT_VECTOR; MAT : TYPE_MATRIX) return BIT_VECTOR is
        variable RESULT : BIT_VECTOR(0 to MAT'length(2)-1);
begin
        assert I'length = MAT'length(1)
            report "Inconsistency between input and product matrix " severity ERROR;
        RESULT := (others => '1');
        for X in MAT'range(2) loop
            for Y in MAT'range(1) loop
                case MAT(X, Y) is
                    when 1   =>  RESULT(X) := RESULT(X) and I(X);
                    when -1  =>  RESULT(X) := RESULT(X) and not I(X);
                    when 0   =>  next;
                end case;
            end loop;
        end loop;
        return RESULT;
end F_AND_PLA;


-- Declaration of the function computing the sums
--      I is the input, i.e. the result of the partial products.
--      MAT is the matrix of sums. Its first index has to be consistent with the output range
--      and Its second index with the number of monoms.
function F_OR_PLA(I : BIT_VECTOR; MAT : TYPE_MATRIX) return BIT_VECTOR is
        variable RESULT : BIT_VECTOR(0 to MAT'length(1)-1);
```

```vhdl
begin
        assert I'length = MAT'length(2)
             report "Inconsistency between output and sum matrix " severity ERROR;
        RESULT := (others => '1');
        for X in MAT'range(1) loop
            for Y in MAT'range(2) loop
                case MAT(X, Y) is
                    when 1   =>   RESULT(X) := RESULT(X) or I(X);
                    when -1  =>   RESULT(X) := RESULT(X) or not I(X);
                    when 0   =>   next;
                end case;
            end loop;
        end loop;
        return RESULT;
end F_OR_PLA;


-- Declaration of the procedure performing the entire computation
procedure P_PLA(   constant M_AND  : TYPE_MATRIX;
                   constant M_OR   : TYPE_MATRIX;
                   signal    I         : in       BIT_VECTOR;
                   signal    O         : out      BIT_VECTOR) is
        variable M : BIT_VECTOR(0 to M_AND'length(2)-1);
begin
        assert M_AND'length(2) = M_OR'length(2)
             report "The number of monoms has to be the same for both matrixes"
             severity ERROR;
        M   :=   F_AND_PLA(I, M_AND);
        O   <=   F_OR_PLA(M, M_OR);
end P_PLA;


constant P        : INTEGER := 4;    -- Number of inputs
constant K        : INTEGER := 3;    -- Number of outputs
constant M        : INTEGER := 6;    -- Number of monoms
type TYPE_PLA is
        record
            AND_MATRIX     : TYPE_MATRIX(0 to P-1, 0 to M-1);
            OR_MATRIX      : TYPE_MATRIX(0 to K-1, 0 to M-1);
        end record;


constant PLA : TYPE_PLA :=
            (   AND_MATRIX    => (   (   1,   -1,   1,   1,   -1,   1),
                                     (   0,   -1,   1,   0,    1,   1),
                                     (   1,   -1,   0,   0,    1,   1),
                                     (   1,    0,   1,   1,    0,   1)  ),
                OR_MATRIX     => (   (   1,   -1,   1,   0,    1,   1),
                                     (  -1,   -1,  -1,  -1,    0,   1),
                                     (   1,   -1,   1,   0,   -1,   1)  )
            );
-- Signals declarations
signal INPUT        : BIT_VECTOR(0 to P-1);
signal OUTPUT       : BIT_VECTOR(0 to K-1);
```

...

```
-- Procedure P_PLA is called from a concurrent context (architecture or block).
LABEL_P_PLA : P_PLA(    M_AND  => PLA.AND_MATRIX,
                        M_OR   => PLA.OR_MATRIX,
                        I      => INPUT,
                        O      => OUTPUT              );
```

**Remarks:**
- The procedure P_PLA may also be called from a sequential context, i.e., from a subprogram or a process.
- The fact of separating product and sum matrices when calling the procedure P_PLA is necessary to ensure the generality of such a subprogram. Indeed, passing the entire structure as a parameter is not possible in the general case: elements of a composite type (array or record) have to be constrained.


## 4.2. SYNCHRONOUS CIRCUITS

Synchronous elements constitute a very large family. It includes latches and registers that are the classical memorization resources. It also involves combinational resources already seen in previous paragraphs: the structure of a synchronous circuit may be considered as a combinational function and a memorization resource (see figure 4.11).

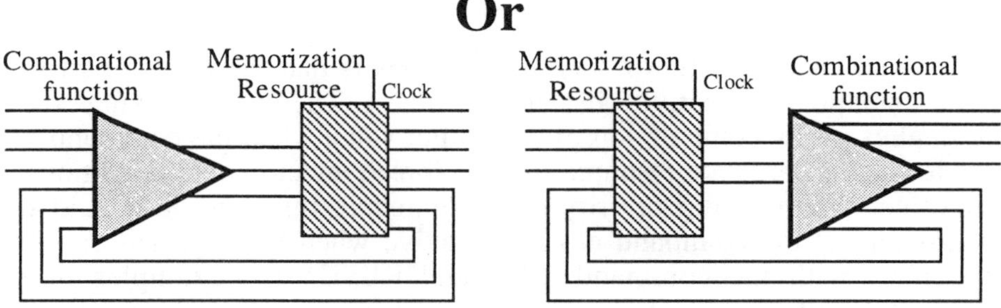


**Figure 4.11.**    **Equivalent Structures of Synchronous Circuits**

Such a resource is said to be sequential, and thus implies a memorization, if the outputs not only depend on the inputs but also on the instant: clock. Latches and flip-flops, which are the basic sequential resources, will be described first. The goal here is to propose a wide variety of possible styles. Thus, deduced from these fundamental models, a set of classical applications involving latches and flip-flops such as counters or finite state machines (FSMs) is presented. Special emphasis is put on FSMs: the VHDL power of description offers a large number of possibilities for their description.


### 4.2.1. Latches

Latches are the simplest memorization resources. Input D is propagated to output Q when the level of command G is active. In the opposite case, the last value of D is present on Q.

Multiple modeling styles are offered for describing latches. The following are among the simplest and most popular.

```
entity LATCH is
  port( G   : in  BIT;
        D   : in  BIT;
        Q   : out BIT );
end LATCH;
```

```
architecture A of LATCH is            -- Architecture B only complies with VHDL'92
begin                                 -- where else clause of concurrent signal
 -- latch on high level               -- assignment is no longer mandatory.
 P : process (G, D)                   architecture B of LATCH is
  begin                               begin
    if G = '1' then                       Q <= D when G = '1';
      Q  <= D;                        end B;
    end if;
   end process P;
end A;
```

The main characteristic of such descriptions is that the data is part of the (explicit or implicit) sensitivity list of the process. Therefore, a change of data value implies a new computation of the output value when the command value allows it. Such a hardware resource may be forced under certain conditions to a given value. When taking priority over command, this forcing is said to be asynchronous. If the command has to be active when forcing, it is said to be synchronous with the command. SET and RESET are examples of such forcings and will be used to illustrate the next examples. Here follows an example of asynchronous SET:

```
entity LATCH is
  port( SET    : in  BIT;
        G      : in  BIT;
        D      : in  BIT;
        Q_S    : out BIT );
end LATCH;
```

```vhdl
architecture ASYNC of LATCH is
begin
  P_LAS : process (G, D, SET)
  begin
    if SET = '1' then
      Q_S    <= '1';
    elsif G = '1' then
      Q_S    <= D;
    end if;
  end process P_LAS;
end ASYNC;
```

The SET signal is part of the sensitivity list of the process and is used first in the process statement. This order is very important and implies the priority of such a signal on the input D and command G of the latch.

Next is the same example but considering the SET signal as synchronous:

```vhdl
architecture SYNC of LATCH is
begin
  P_LAS : process (G, D)
  begin
    if G = '1' then
        if SET = '1' then
            Q_S       <= '1';
        else
            Q_S       <= D;
        end if;
    end if;
  end process P_LAS;
end SYNC;
```

Here the signal SET has no longer to be in the sensitivity list. Its value is only read when the input G value is high.

As illustrated in the following example, the level of abstraction of datas, as well as those of commands, is not *a priori* restricted. Taking again an asynchronous example, it is possible to write the following:

```vhdl
type TYPE_DATA is (SOUND, IMAGE, TEXT);
signal DATA, DATA_1  : TYPE_DATA;
signal CMD                 : INTEGER range 0 to 23;

...                                                    -- Second version
-- First version (VHDL'92)                             process (CMD, DATA_1)
DATA  <=  TEXT          when CMD = 0     else          begin
          DATA_1        when CMD >= 16   else              if CMD = 0 then
          unaffected;                                          DATA  <=  TEXT;
                                                           elsif CMD = 16 then
                                                               DATA  <=  DATA_1;
                                                           end if;
                                                       end process;
```

This is equivalent to previously seen modeling styles as shown by the following transformation. Indeed, CMD = 0 may be considered to be a set command, and CMD=16 the gate (signal G).

```
...
type TYPE_DATA is (SOUND, IMAGE, TEXT);
signal DATA, DATA_1   :  TYPE_DATA;
signal CMD                 : INTEGER range 0 to 23;
signal G, SET              : BOOLEAN;
...
G      <= (CMD = 16);
SET   <= (CMD = 0);
...
-- First version (VHDL'92)
DATA  <=  TEXT          when SET     else
              DATA_1       when G        else
              unaffected;

-- Second version
process (SET, G, DATA_1)
begin
      if SET then
            DATA  <=  TEXT;
      elsif G then
            DATA  <=  DATA_1;
      end if;
end process;
```

For a good efficiency of the design, latch inferences have to be carefully checked by the designer. Indeed, the difference in VHDL code between a latch and a multiplexer is small. In the case of latches, the output is simply not assigned in one branch.

In the general case, if at least one signal whose value is read within a process is not part of the sensitivity list, a latch resource is inferred.

## 4.2.2. Registers

The notion of register is similar to that of latch. Data are memorized on the edge of the command instead of its level. Descriptions of latches and registers are therefore very close. Taking again the example of latches, it is easily possible to change them into registers. The only modification consists of making the process insensitive to events on datas.

```
entity FF is
  port( CLK      : in  BIT;
        D    : in   BIT;
        Q    : out BIT );
end FF;


architecture A of FF is
begin
```

```
P_FF : process                                  P_FF : process (CLK)
begin                                           begin
        wait until CLK = '1';        <==>           if CLK = '1' then
        Q <= D;                                         Q <= D;
                                                    end if;
    end process P_FF;                           end process P_FF;
end A;
```

Two ways of describing registers have been seen above. Many other modeling styles are possible. Using a concurrent signal assignment, the following description illustrates one of them:

```
signal Q_LOCAL : BIT;  -- Since output signal cannot be read, this declaration is needed
...
Q_LOCAL <= D when CLK and CLK'event else Q_LOCAL;
Q <= Q_LOCAL;
```

*VHDL'92* offers other possibilities:

```
-- D and Q are signals of type BIT. CLK is a clock signal of type  BOOLEAN.
Q <=   D when CLK and CLK'event else Q'DRIVING_VALUE;   -- VHDL'92 compliant
Q <=   D when CLK and CLK'event else unaffected ;       -- VHDL'92 compliant
Q <=   D when CLK and CLK'event ;                        -- VHDL'92 compliant
```

It is important to note that the previous use of attribute 'EVENT is here a necessity and not a simple convention for synthesis tools. Indeed, the equivalent process of the signal assignment is sensitive to any event on signal D or CLK. The purpose of the condition "**and** CLK'event" is to filter the events on D. If these events were taken into account, the description would be a latch instead of a register: any change on the data would imply a change on the output when CLK is true.

As shown in the following example, a specific feature of VHDL — guarded blocks — may also be used for describing registers.

```
-- D and Q are signals of type INTEGER. CLK is a clock signal of  type BIT.
B_FF : block (CLK='1' and not CLK'stable)
begin
        Q <= guarded D;
end block B_FF;
```

Here again, the use of the expression "**and not** CLK'STABLE" is necessary to obtain the functionality of a register. It is important to note that the use of the attribute 'EVENT is not possible in this case, as the model would become a latch. This is not obvious at first glance, but it has been already discussed in paragraph 3.4.3. and constitutes a real trap for designers.

As for latches, registers may be forced to be synchronous or asynchronous. Starting from latch descriptions and taking care to remove the input from the sensitivity list, register descriptions may be obtained. Here follows an entity with two architectures — synchronous and asynchronous — modeling a register.

```vhdl
entity FF is
 port( RESET   : in  BIT;
       CLK     : in  BIT;
       D       : in  BIT;
       Q       : out BIT );
end FF;

architecture ASYNC of FF is
begin
 P : process (CLK, RESET)
 begin
   if RESET = '0' then
     Q <= '1';
   elsif CLK = '1' and CLK'event then
     Q <= D;
   end if;
  end process P;
end ASYNC;

architecture SYNC of FF is
begin
 P : process (RESET, CLK)
 begin
   if CLK = '1' and CLK'event then
       if RESET = '0' then
           Q    <= '1';
       else
           Q    <= D;
       end if;
    end if;
  end process P;
end SYNC;
```

The asynchronous RESET is part of the sensitivity list and is read first in the process statement to give it the higher priority.

In the previous synchronous architecture, a combinational hardware resource may be extracted: signal Q, depending on the value of RESET, receives either '1' or the current value of input signal D. As already studied in paragraph 4.1.5., the inferred hardware resource is a multiplexer. As shown in figure 4.12, this multiplexer is simply used in a synchronous context:

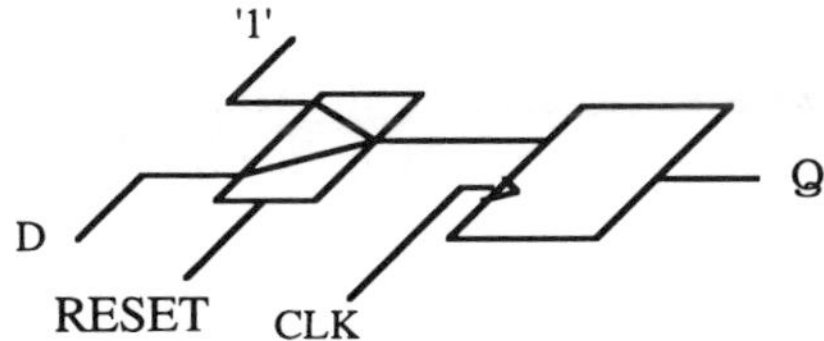


**Figure  4.12.        Multiplex  in  Synchronous  Context**

Starting from this last example, the description of a register with a command enabling the memorization may be easily deduced:

```
-- ENABLE is the synchronous command of type boolean allowing the updates.
-- Q is the output signal, D is the input and CLK is a clock signal of type BIT.
P_ENABLE : process
begin
        wait until CLK = '1';
        if ENABLE then
            Q <= D;
        end if;
end process P_ENABLE;
```

The functionality here is very close to a synchronous forcings, previously illustrated with synchronous set and reset. The main difference is that the forcing command (ENABLE) does not imply a given predefined value to be forced, but the previously memorized value itself (Q).

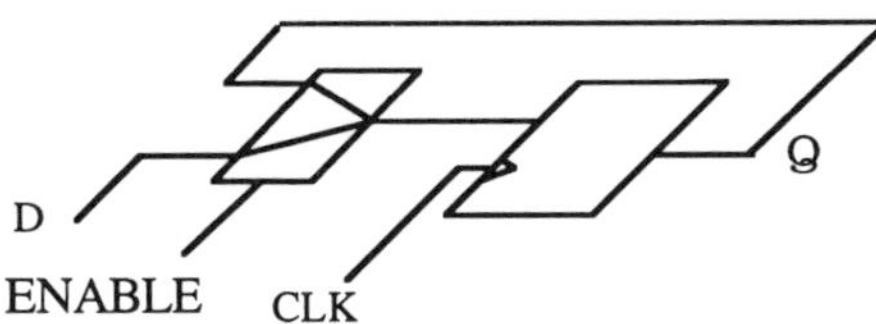


**Figure  4.13.        Register  with  Synchronous  Enable**

Indeed, memorization resources may be described using explicit combinational function. A general form of this modeling style is

```
process
begin
        wait until CLK = '1';
        Q <= F (D, Q, ...);
end process;
```

Function F may express a classical Boolean equation as well as an arithmetic operation. This modeling style is concise, and thus improves the readability of the description.

**Remark:**
*VHDL'92* makes a distinction between pure and impure functions (possibly involving memorization). Function F has to be pure, as are all *VHDL'87* functions.

The following example applies this modeling style to addition operations:

```
-- A, B, S, Q are signals of the same type (or at least compatible).
-- CLK is a clock signal of type BIT.

-- First option: separate synchronous and combinational descriptions.
S <= A + B;
P_CLK : process
begin
        wait until CLK = '1';
        Q <= S;
end process P_CLK;

-- Second option: associate synchronous and combinational descriptions in the same process.
P_CLK : process
begin
        wait until CLK = '1';
        Q <= A + B;
end process P_CLK;
```

## 4.2.3. Synchronous Counters

Synchronous counters, which constitute a large family, are a good illustration of the principle given by figure 4.11. Any combinatory function may be used with a synchronous context if memorization is needed.

Counters have been overviewed in a previous section (4.1), dedicated to combinational hardware. The most simple example of counter description in a synchronous context is a counter modulo 2**P:

```
-- P is a constant of type NATURAL.
-- C is a signal of subtype INTEGER range 0 to 2**P-1.
-- CK is a clock signal of type BIT.
P_COUNT : process
begin
        wait until CK = '1';
        C <= (C + 1) mod 2**P;
end process P_COUNT;
```

Unfortunately, this style is rarely accepted by synthesis tools. It is therefore more reasonable to prefer the following description:

```
-- P is a constant of type NATURAL.
-- C is a signal of type INTEGER range 0 to 2**P-1.
-- CK is a clock signal of  BIT.
P_COUNT : process
begin
        wait until CK = '1';
        if C = (2**P)-1 then
            C <= 0;
        else
            C <= C + 1;
        end if;
end process P_COUNT;
```

Another equivalent description consists in using a bit-array type for the data. The standard numerical package (NUMERIC_BIT) offers such a type (UNSIGNED) as well as the usual arithmetic operations to be performed on it (operator overloading). Therefore, the description becomes straightforward:

```
library IEEE;
use IEEE.NUMERIC_BIT.all;      -- Reference to the standard arithmetic package
...
...
-- P is a constant of type NATURAL.
-- C is a signal of type UNSIGNED(P-1 downto 0).
-- CK is a clock signal of type BIT.
P_COUNT : process
begin
        wait until CK = '1';
        C <= C + 1;
end process P_COUNT;
```

Unfortunately, if only the modulo parameter (let say N) is known (and is possibly different from a power of two, 2**P), the number of bits strictly necessary to code N-1 has to be computed. This is a simple logarithm in base two, but this computation makes the description more complex:

```
library IEEE;
use IEEE.NUMERIC_BIT.all;      -- Reference to the standard arithmetic package
use IEEE.MATH_REAL.all ;       -- Reference to a mathematic package
...
...
-- N is a constant of type NATURAL.
-- C is a signal of type UNSIGNED(INTEGER(LOG(2, REAL(N)))-1 downto 0).
-- CK is a clock signal of type BIT.
```

```vhdl
P_COUNT : process
begin
      wait until CK = '1';
      if TO_INTEGER(C) = N-1 then
          C <= (others => '0');
      else
          C <= C + '1';          -- Overloaded addition between UNSIGNED and BIT types
      end if;
end process P_COUNT;
```

Coming back to the other style, a type INTEGER may be used to represent the data. The resulting VHDL code is thus

```vhdl
-- N is a constant of type NATURAL.
-- C is a signal of type INTEGER range 0 to N-1.
-- CK is a clock signal of  BIT.
P_COUNT : process
begin
      wait until CK = '1';
      if C = N-1 then
          C <= 0;
      else
          C <= C + 1;
      end if;
end process P_COUNT;
```

or the following code:

```vhdl
-- N is a constant of type NATURAL. C is a signal of type INTEGER range 0 to N-1.
-- CK is a clock signal of  BIT.
P_COUNT : process
begin
      wait until CK = '1';
      C <= C + 1;
      if C = N-1 then
          C <= 0;
      end if;
end process P_COUNT;
```

**Remark:**

In the two previous examples, a signal is used to store the counter value. Although it is quite easy to modify the first description to use a variable instead of a signal, modifying the second one leads to a trap. The test of the old counter value is performed after the assignment of the new value. If a variable is used instead of a signal, since the target of a variable assignment is immediately updated, the function carried out will be different and another memorization necessary to store the incremented value.

The following example, chosen because it is more complex than the previous ones, illustrates the initialization issues. As shown by figure 4.14, the value of the increment of this counter depends on the value of the input. Only positive values are shown in this figure, but this function is assumed to be symmetrical with respect to the origin.

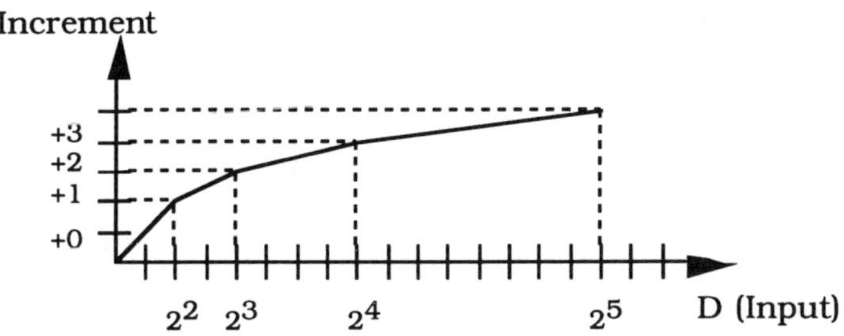


**Figure 4.14.** **Logarithmic Conversion Function**

Data are considered periodical but no clock signal is explicitly given. VHDL offers an attribute ('TRANSACTION) to handle transactions on such data: each time the data are assigned, even when not modified, a transaction occurs. In this example, the input signal D is considered to be emitted by a synchronous source. Its implicit clock may be recovered using a synchronization statement (**wait** statement) on the attribute of kind signal D'TRANSACTION.

**Remark:**
This notion of transaction on an input is only an abstract concept. Indeed, inferred hardware requires a clock.

```
-- Within a process...
wait on D'transaction;
-- Instead of the well-known:
wait until CLK = '1';                    -- CLK is a signal of type BIT

-- Or if the synchronization is analyzed in an expression (within a process)
if D'active then ...
-- Instead of:
if CLK = '1' and CLK'event then ...      -- CLK de type BIT
```

The counter is either incremented or decremented (negative values of D) depending on the value of D (see figure 4.14). The range of data is determined by the value of the generic N and the range of the counter is given by the value of generic P. A RESET input allows for an asynchronous rest of the counter.

```vhdl
entity LOG_CODAGE is
        generic   (    N :   POSITIVE := 5;
                       P :   POSITIVE := 4      );
        port      (    RESET : in BIT;
                       D    : in   INTEGER range -2**(N-1)+1 to 2**(N-1) - 1;
                       C    : out INTEGER range -2**(P-1) to 2**(P-1) - 1    );
end;

architecture A of LOG_CODAGE is
        -- This local signal duplicates the port C (output port) in order to allow its reading.
        -- VHDL'92 makes this declaration useless: C'driving_value may be read.
        signal S_C : INTEGER range -2**(P-1) to 2**(P-1) - 1;
begin
        C <= S_C;
        MAIN_P : process (RESET, D'transaction)
            variable V     : INTEGER range 0 to 2**(N-1)-1;
            variable INC   : INTEGER range 0 to 2**(P-1)-1;
        begin
            if RESET = '1' then
                C <= 0;
            elsif D'active then
                -- Computation of the absolute value of D
                if D < 0 then
                    V := -D;
                else
                    V := D;
                end if;
                -- Computation of the increment
                INC := 0;
                for I in 2 to N loop
                    if V >= 2**(I-1) then
                        INC := I-2;
                        exit;
                    end if;
                end loop;
                -- Counter update
                if D < 0 then
                    S_C <= S_C - INC;
                else
                    S_C <= S_C + INC;
                end if;
            end if;
        end process MAIN_P;
end A;
```

When writing such a description, it is perfectly possible to predict which hardware resources are necessary. Indeed, the designer *has* to do so: writing synthesizable code means carefully controling the function carried out, and also the quality and efficiency of the inferred hardware.

For example, the operation D < 0 is used twice and has therefore to be shared. The operation of addition or subtraction, depending on the sign of D, may be shared as a single adder-subtractor whose use is exclusive in time. Finally, the designer has to know the exact number of inferred registers. In this example, only signal S_C has to be memorized. All local variables are only used for intermediate computations of the increment. Indeed, it is perfectly possible to envisage such a treatment as a function. Thus, its combinational nature appears obvious (cf. paragraph 3.4.6.).

The following example of a multiphase clock generator closes this section. This kind of hardware resource is useful when a periodic and complex sequencing is required.

Figure 4.15 defines the specifications of such a device: the input is a single clock and the outputs are other clocks whose period is a multiple of the input clock period. Moreover, these clocks are not necessarily based on the same edge (rising for PH1 and PH2, falling for PH3).

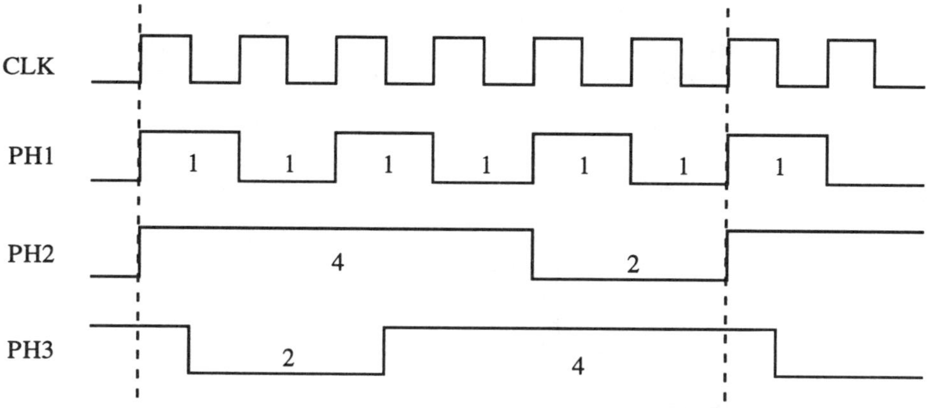


**Figure 4.15.** **Multiphase Clock Generator**

Outputs have the same global period. In our example, the global period involves six input clock periods. This duration falls between the two vertical dotted lines. The modeling strategy is therefore to create a counter (or decounter) modulo 6 and to determine the values of outputs with respect to the internal state of this counter..

Output PH3 generation is somewhat specific due to the fact that this signal is sensitive to falling edges of CLK. Since a majority of signals (PH1 and PH2) are based on rising edges of CLK, a signal PH3_1 is generated on the rising edge of CLK. From this signal PH3_1, signal PH3 is therefore easily deduced by shifting it with half a period of CLK.

Here follows the corresponding description:

```
entity CLOCK is
  port (    RESET  : in  BIT;             -- asynchronous reset
            CLK    : in  BIT;             -- reference clock
            PH1    : out    BIT;          -- active on rising edge of CLK
            PH2    : out    BIT;          -- active on rising edge of CLK
            PH3    : out    BIT);         -- active on falling edge of CLK
end CLOCK;

architecture A of CLOCK is
      signal PH3_R : BIT; -- PH3 is synchronous with the rising edge of CLK
begin
      P1 : process
          variable K : INTEGER range 0 to 5;
      begin
          if RESET = '1' then
                K := 0;
          elsif CLK = '1' and CLK'event then
                case K is
                      when 0   =>  PH1 <= '1';    PH2 <= '1';    PH3_R <= '0';
                                   K := 5;
                      when 1   =>  PH1 <= '0';    PH2 <= '1';    PH3_R <= '0';
                                   K := K - 1;
                      when 2   =>  PH1 <= '1';    PH2 <= '1';    PH3_R <= '1';
                                   K := K - 1;
                      when 3   =>  PH1 <= '0';    PH2 <= '1';    PH3_R <= '1';
                                   K := K - 1;
                      when 4   =>  PH1 <= '1';    PH2 <= '0';    PH3_R <= '1';
                                   K := K - 1;
                      when 5   =>  PH1 <= '0';    PH2 <= '0';    PH3_R <= '1';
                                   K := K - 1;
                end case;
            end if;
            wait on RESET, CLK;
      end process P1;

      P2 : process               -- Half-period delay
      begin
          wait until CLK = '0';
          PH3 <= PH3_R;
      end process P2;

end A;
```

Here again, the designer has to think it out: how many memorization resources are necessary, and what is their nature?

Indeed, each signal and each variable (which is not a simple intermediate variable for computation) assigned in a synchronous context infers a memorization.

Architecture A of entity CLOCK involves two synchronous processes.

Process P1 is sensitive to rising edges of CLK and P2 is sensitive to the corresponding falling edges. Thus, memorization elements, which have to be active on edges, are flip-flops.

The inferred flip-flops are

- Three flip-flops sensitive to rising edges are necessary to memorize the value of the counter K (0 to 5) of process P1. Each of them possesses an asynchronous reset.
- Three flip-flops sensitive to rising edges, are also necessary for signals PH1, PH2 et PH3_R.
- One flip-flop, active on a falling edge, is used for signal PH3.

Once these resource memorizations are known, the designer has to answer the question: What is strictly necessary?

Flip-flops associated with the counter as well as this one synchronizing PH3 on the falling edge of CLK are obviously necessary. This is not the case of the flip-flops memorizing the output values of PH1, PH2 and PH3_R. Indeed, these outputs may be computed using combinational functions from the counter value. Therefore, the following architecture may seem more efficient.

```
architecture B of CLOCK is
        signal K : INTEGER range 0 to 5;
        signal PH3_R : BIT;                   -- PH3 is synchronized on the rising edge of CLK
begin

        PK : process
        begin
            if RESET = '1' then
                K <= 0;
            elsif CLK = '1' and CLK'event then
                K <= K - 1;
            end if;
            wait on RESET, CLK;
        end process PK;

        P1 : process
        begin
            case K is
                when 0  =>  PH1 <= '1';   PH2 <= '1';   PH3_R <= '0';
                when 1  =>  PH1 <= '0';   PH2 <= '1';   PH3_R <= '0';
                when 2  =>  PH1 <= '1';   PH2 <= '1';   PH3_R <= '1';
                when 3  =>  PH1 <= '0';   PH2 <= '1';   PH3_R <= '1';
                when 4  =>  PH1 <= '1';   PH2 <= '0';   PH3_R <= '1';
                when 5  =>  PH1 <= '0';   PH2 <= '0';   PH3_R <= '1';
            end case;
            wait on K;
        end process P1;
```

```
P2 : process                    -- Shift of half a period
begin
        wait until CLK = '0';
        PH3 <= PH3_R;
    end process P2;
end B;
```

Obviously, this second architecture saves three flip-flops. However, given the context, is it the right description? Since the goal of this model is to generate clocks, it is therefore essential to prohibit glitches on the outputs. Unfortunately, architecture B has some drawbacks on this point. Process P1 is combinational and sensitive to K. The designer has to remember that the hardware representation of an integer is a bit vector. When K is decremented from four to three, many events appear on the bit vector representation, from "100" to "011", the three bits have to switch. The changes, due to the fact that P1 is combinational, are immediately propagated on the outputs, creating glitches as illustrated in figure 4.16. This is not acceptable for a clock generator. Architecture A, although more expensive in terms of flip-flops, avoids this problem. Outputs are only computed on a stabilized value of K.

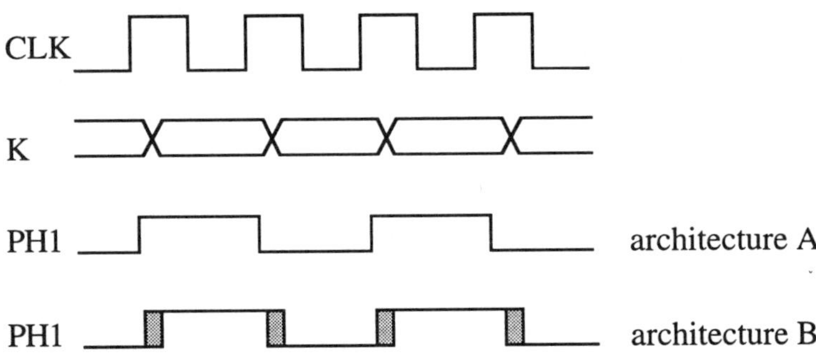


**Figure  4.16.**          **Comparison Between Architectures A and B**

This example does not imply that many registers have to be used systematically. The conclusion is quite different. For an efficient result, the designer has to deeply analyze the nature (combinational or sequential) of all signals or variables.


### 4.2.4. Memories

In this paragraph, the principal organizations of memorization elements are presented. Such organizations are usually only envisaged for a reasonable memory size. If a considerable number of elements are needed, other solutions are possible, such as characterized RAMs for ASICs or external memorization resources.

A first example of such an organization, a register file of 32 words of 8 bits, is now given. The control of this register file is handled by an address bus AD, a read (RD) command, and a write (WR) command. All these signals are synchronized on a clock (CLK). Register file data are read or written through the bidirectional bus DATA. Since this bus is shared with other register files, its type is resolved: STD_LOGIC_VECTOR.

**Figure  4.17.**        **External View of the Register File**

A VHDL description of the register file is

```vhdl
library IEEE;
use IEEE.STD_LOGIC_1164.all;
use IEEE.NUMERIC_STD.all;

entity REGISTER_FILE is
   port (   CLK     : in  STD_ULOGIC;
            RD      : in  STD_ULOGIC;
            WR      : in  STD_ULOGIC;
            AD      : in  UNSIGNED(4 downto 0);
            DATA    : inout   STD_LOGIC_VECTOR(7 downto 0));
end REGISTER_FILE ;

architecture A of REGISTER_FILE is
begin
      P : process
         type TYPE_M is array (NATURAL range 0 to 2**AD'length-1)
               of  STD_LOGIC_VECTOR(DATA'range);
         variable M : TYPE_M;
      begin
            wait until CLK = '1';
            -- by default the bus has to be set to 'Z'
            DATA <= (others => 'Z');
            if WR = '1' then
                  M(TO_INTEGER(AD)) := DATA;
```

```
        elsif  RD = '1' then
             DATA  <=  M(TO_INTEGER(AD)) ;
        end if;
     end process P;
  end A;
```

The function TO_INTEGER converts an unsigned vector into a natural. This function is exported by the standard arithmetic package NUMERIC_STD. This package is presented in section 6.3.

The designer has to be careful that a 'Z' value is assigned to all bus elements when this bus is (theoretically) not driven.

The following example deals with shift registers. Descriptions of such hardware resources have already been studied in paragraph 4.1.4. Shift registers are often used when modeling synchronous systems. They are useful as soon as it is necessary to delay from a phase clock, and in a more general case, if pipelining is used.

This first example of shift register is parametrized according to the desired depth of the pipelining (which can be zero) and the size of the words. If there is no pipelining, output DOUT simply reproduces the input DIN.

```
entity PIPE is
  generic ( N    : POSITIVE;              -- size of words
            K    : NATURAL  );            -- pipelining depth
     port (    CLK     : in      BIT;
               DIN     : in      BIT_VECTOR(N-1 downto 0);
               DOUT    : out     BIT_VECTOR(N-1 downto 0)   );
end PIPE;

architecture A of PIPE is
begin
       B_K :  if K >= 2 generate
            B_LOCAL : block
                 type TYPE_R is array (NATURAL range 0 to K-1)
                       of BIT_VECTOR(N-1 downto 0);
                 signal R : TYPE_R;
            begin
                 B1 : for I in 0 to K-2 generate
                     process
                     begin
                          wait until CLK = '1';
                          R(I+1) <= R(I);
                     end process;
                 end generate B1;
                 R(0)      <= DIN;
                 DOUT    <= R(K-1);
            end block B_LOCAL;
       end generate B_K;
```

```
B_K_one :  if K =1 generate
     process
     begin
             wait until CLK = '1';
         DOUT <= DIN;
     end process;
end generate B_K_one;


B_K_zero :  if K = 0 generate
     DOUT <= DIN;
end generate B_K_zero;


end A;
```

The previous description is *VHDL'87* compliant. Using new *VHDL'92* new constructs, certain simplifications are possible. For example, the subblock B_LOCAL of block B_K may be suppressed. Its only purpose is to define a declarative part, local to the **generate** statement. The *VHDL'92* version of the **generate** statement optionally includes such a declarative part.

Another kind of memorization resource, stacks, is now studied. In a first step, a synchronous LIFO (Last In First Out) description is proposed (cf. figure 4.18). When signal PUSH is active, the value of signal DIN (data) is stored at the top of the stack. Other elements of the stack are shifted. Signal DOUT returns this value when signal POP is active. Other elements are then shifted back: the second element is put on top, the third becomes the second, and so on.

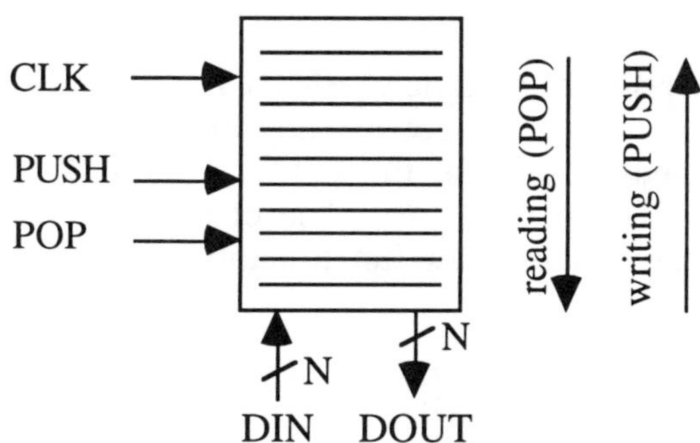


**Figure 4.18.**        **External View of a LIFO**

This behavior may be described using the following model:

```vhdl
entity LIFO is
  generic ( N    : POSITIVE;        -- Size of an element (number of bits)
            K    : POSITIVE );      -- Size of the stack (number of elements)
    port (    CLK     : in      BIT;
              PUSH    : in      BIT;
              POP     : in      BIT;
              DIN     : in      BIT_VECTOR(N-1 downto 0);
              DOUT    : out     BIT_VECTOR(N-1 downto 0)    );
end LIFO;

architecture A of LIFO is
begin
        process
            type TYPE_R is array (NATURAL range K-1 downto 0)
                         of BIT_VECTOR(N-1 downto 0);
            variable R : TYPE_R;
        begin
            wait until CLK = '1';
            if PUSH = '1' then
                R(K-1 downto 1) := R(K-2 downto 0);
                R(0) := DIN;
            elsif POP = '1' then
                DOUT <= R(0);
                R(K-2 downto 0) := R(K-1 downto 1);
            end if;
        end process;

    end A;
```

Starting from this simple model, it can be interesting to add signals to check the states of the stack. For example, specific outputs may indicate that the LIFO is full or empty. This implies counting the number of reading (POP) and writing (PUSH) operations.

For this purpose, two signals, EMPTY and FULL, may be added to the entity declaration. To ensure consistency between the internal counter and the number of elements stored in the LIFO, a RESET input that initializes the counter to zero is also necessary. The changes to be made on the previous architecture to handle these new functionalities are

```vhdl
-- To be added to the entity ports
          RESET   : in      BIT;
          EMPTY   : out     BIT;
          FULL    : out     BIT;


-- To be added to the declarative part of architecture A
      signal C : INTEGER range 0 to K-1;
      function TO_BIT (B : in BOOLEAN) return BIT is
      begin
```

```
              case B is
                  when TRUE   => return '1';
                  when FALSE  => return '0';
              end case;
          end TO_BIT;

  -- To be added to architecture A
          EMPTY          <= TO_BIT(C= 0);
          FULL           <= TO_BIT(C= K-1);

          process (RESET, CLK)
          begin
              if RESET = '1' then
                  C <= 0;
              elsif CLK = '1' and CLK'EVENT then
                  if PUSH = '1' then
                      C <= C + 1;
                  elsif POP = '1' then
                      C <= C - 1;
                  end if;
              end if;
          end process;
```

Indeed, it is perfectly possible to put together in the same process, the stack and counter managements.

It is important to note that ports EMPTY and FULL are updated outside synchronous processes. This allows memorization elements (flip-flops) to be saved. Indeed, a memorization is not really necessary in their case. Although glitches may appear on signals EMPTY and FULL in the previous description, the hypothesis made here is that their value is only read from another synchronous context based on the same clock CLK, and therefore their state is stabilized during reading.


## 4.2.5. Finite State Machines

Finite State Machines allow for description of complex controls. This modeling technique has been wellknown for a long time and is very popular. VHDL, although not providing a dedicated "finite state machine statement," offers instead a great number of modeling possibilities.

Indeed, a large part of VHDL's power of description comes from its ability to describe any kind of finite state machine. Among other VHDL features, all the capabilities of the set of VHDL sequential statements are available for such descriptions. As in other domains when synthesis is the goal, it is nevertheless necessary to reduce the variety of modeling styles. Nowadays, synchronous finite state machines are the only ones to be accepted by synthesis tools.

The purpose of this section is not to provide a course on automatons in their generality: it is even not important for the designer to know if his or her result will be a Mealy or Moore machine. What is really needed is a perfect consistency of behavior between the automaton as simulated and as synthesized.

Some basic characteristics of finite state machines may be summarized as follows:

- The automaton is always in one of its possible states: the current state. This state is memorized in a specific register — state register.
- The next state may be computed using the current state and the input values.
- Output values are computed depending on either the current state or the transition between two states.
- During each clock period, the state register is updated with the previously computed state (next state).

**Remark:**

Since a state is always computed using the previous state, it is essential to control the value of the initial state. For this purpose, synchronous or asynchronous initialization mechanisms may be used:

The template of a finite state machine description may be the following:

```
-- RESET is the asynchronous reset, CLK is the clock and STATE
-- is a variable (or a signal) memorizing the current state.
process (RESET, CLK)
begin
        if RESET then
            STATE  ...
        elsif CLK = '1' and CLK'EVENT then
            STATE  ...
        end if;
end process;
```

This modeling style is similar to those used in paragraph 4.2.2. for registers. An example using the previous template is now given. The goal of the model is to provide a periodic output whose period is constant but whose clock-level ratio may change (cf. figure 4.19).

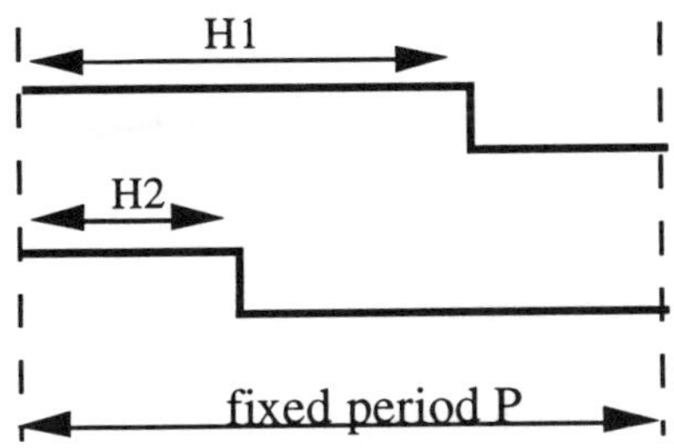


**Figure 4.19.**       **Same Period P, Different Clock-Level Ratios: H1 and H2**

The computation of the fixed period P as well as the duration H of the high level are based on counting a faster clock.

Reasoning in the context of finite state machines, the proposed behavior may be decomposed into several phases (or states). These phases are repeated indefinitely. The number of different phases is simple to determine. Only two decisions have to be taken: assigning the output to '1' or deciding to put it to '0'. Between these two decisions, the only operation consists in counting the clock period. Two states are therefore necessary to model this behavior:

First state
- If signal H is equal to zero, output O takes the value '0' and counting is started to wait for the end of the period. The second state will not be reached during this period.
- If signal H is equal to P, the treatment is similar to the previous one with the exception that output O takes the value '1'.
- For other values of signal H, output O is set to '1' and a counting is started and decides the end of the high level of the output. The value of H is memorized.

Second state
- When the end of the high level duration is reached, output O is set to '0' and counting is started to decide the end of the period. Of course, the end of this period is also the beginning of the next one.

Here follows the VHDL description of such a device:

```vhdl
entity RATIO is
   generic (P : POSITIVE);        -- Period expressed as a number of CLK periods.
   port (   RESET  : in      BIT;
            CLK    : in      BIT;
            H      : in      INTEGER range 0 to P-1;
            O      : out     BIT                    );
end RATIO;
```

```vhdl
architecture A of RATIO is
begin
 P_FSM : process (CLK, RESET)
      type STATES is (S1, S2);
      variable STATE   : STATES;
      variable K           : INTEGER range 0 to P-1;
      -- In order to memorize H during the period.
      variable M_H         : INTEGER range 0 to P-1;
  begin
      if RESET = '1' then
          K := 0;
          STATE := S1;
      elsif CLK = '1' and CLK'event then
          case STATE is
              when S1 =>   if K = 0 then
                               M_H := H;                    -- Memorization of the value of H
                               if H = 0 then
                                   O    <= '0';
                                   K    := P-1;
                                   STATE := S1;
                               elsif H = P-1 then
                                   O    <= '1';
                                   K    := P-1;
                                   STATE := S1;
                               elsif H = 1 then
                                   O    <= '1';
                                   K    := 0;
                                   STATE := S2;
                               end if;
                           else
                               K := K - 1;
                           end if;
              when S2 =>   if K = 0 then
                               O    <= '0';
                               K    := P - M_H -1;
                               STATE := S1;
                           else
                               K := K - 1;
                           end if;
          end case;
      end if;
  end process P_FSM;

end A;
```

As has become usual in this chapter, a quick list of necessary hardware
resources must be made. Counter K implies (P-1) bits of memory, the state
register only one bit (two states S1 and S2 have to be encoded), and output O
requires another bit. Furthermore, memory elements associated with counter K

and the state register have to offer an asynchronous initialization command (when RESET is equal to '1').

Since it is sometimes desirable to control, at least for outputs, the inferred flip-flops, the designer has to modify his or her modeling style. The fact that outputs are assigned in the context of a synchronous automaton implicitly implies their memorization. To avoid this, output assignments have to be separated from the state updates. The state update operation may itself be decomposed into two treatments: next state computation and current state update.

In the previous example (architecture A), the automaton is described in the most concise way. A unique process performs the three treatments: next state computation, current state update and output assignments. These inferred hardware elements are expensive; a register or flip-flop is used for each output. When this modeling style is used, since all treatment is local to a process, the current state may be either a signal or a variable.

Other ways of modeling automatons consist of grouping the three treatments differently. Architecture B proposes to split these treatments into three different processes:

```
library IEEE;
use IEEE.STD_LOGIC_1164.all;
entity X is
    port (  RESET          : in      STD_ULOGIC;
            CLK            : in      STD_ULOGIC;
            I0, I1, ..., Ik : in     STD_ULOGIC;
            O0, O1, ..., Op : out    STD_ULOGIC    );
end X;

architecture B of X is
        type TYPE_STATE is (S_INIT, S1, S2, ..., Sn);
        signal CURRENT_STATE : TYPE_STATE;
        signal NEXT_STATE    : TYPE_STATE;
begin
        P_STATE : process
        begin
            wait until CLK = '1';
            if RESET = '1' then
                CURRENT_STATE <= S_INIT;
            else
                CURRENT_STATE <= NEXT_STATE;
            end if;
        end process P_STATE;

        P_NEXT_STATE : process
        begin
            -- Assignment by default
            NEXT_STATE <= CURRENT_STATE;
            case CURRENT_STATE is
```

```
            when S_INIT  =>   if (I0 and I1) = '1' then
                                    NEXT_STATE <= S5;
                              elsif I7 = '0' then
                                    NEXT_STATE <= S2;
                              end if;
            when S1          =>  if I8 = '1' or I3 = '0' then
                                    NEXT_STATE <= S4;
                              end if;
            when Sn          => ...
        end case;
        wait on I0, I1, ..., Ik, CURRENT_STATE;
    end process P_NEXT_STATE;


    P_OUTPUTS: process
    begin
        -- Assignments by default
        O0  <= '0';
        O1  <= '1';
        ...
        case CURRENT_STATE is
            when S_INIT     => O4 <= '0';
                               O5 <= '0';
            when S1         => O2 <= '1';
            when Sn         => ...
        end case;
        wait on CURRENT_STATE;
    end process P_OUPUTS;

end B;
```

From these three processes, only one (P_STATE) is sequential. It infers the state register of the finite state machine. The initialization of this register is synchronous.

To simplify the writing of combinational treatments, next state computation and output values, it is interesting to use the notion of default value. It consists in systematically assigning a signal to the value it has in the majority of cases. This is done before **if** or **case** statements. This statement is therefore much simpler since it only involves particular case assignments.

From a modeling point of view, the previous automaton is described as a Moore machine: the output values only depend on the current state. It is easy to change it into a Mealy description by merging the two combinational processes (P_NEXT_STATE and P_OUTPUTS) and by assigning the outputs on the state transitions. Switching from one description to the other may be done in the following ways:

- by keeping process P_STATE and dispatching output assignments into the two other processes, which become

```vhdl
P_NEXT_STATE : process
begin
    -- Default assignment (necessary to have a combinational process)
    NEXT_STATE <= CURRENT_STATE;
    -- Default values for outputs assigned in this process
    O3 <= '0';
    case CURRENT_STATE is
        when S_INIT  =>  if (I0 and I1) = '1' then
                             O6 <= '1';
                             NEXT_STATE <= S5;
                         elsif I7 = '0' then
                             O6 <= '0';
                             NEXT_STATE <= S2;
                         end if;
        when S1          =>  if I8 = '1' or I3 = '0' then
                             O3 <= '1';
                             NEXT_STATE <= S4;
                         end if;
        when Sn          => ...
    end case;
    wait on I0, I1, ..., Ik, CURRENT_STATE;
end process P_NEXT_STATE;


P_OUTPUTS: process
begin
    -- Default values for outputs assigned in this process
    O0  <= '0';
    O1  <= '1';
    ...
    case CURRENT_STATE is
        when S_INIT     => O4  <= '0';
                           O5  <= '0';
        when S1          => O2  <= '1';
        when Sn          => ...
    end case;
    wait on  CURRENT_STATE;
end process P_OUTPUTS;
```

- by keeping process P_STATE, merging the two other processes, and assigning the outputs conditionally to the state or state transition.

```vhdl
P_NEXT_STATE : process
begin
    -- Default assignment (necessary to have a combinational process)
    NEXT_STATE <= CURRENT_STATE;
    -- Default values for outputs assigned in this process
    O0  <= '0';
    O1  <= '1';
    O3  <= '0';
    case CURRENT_STATE is
```

```
        when S_INIT      => O4  <= '0';          -- on state
                            O5  <= '0';
                            if (I0 and I1) = '1' then
                                O6 <= '1';      -- on state transition
                                NEXT_STATE <= S5;
                            elsif I7 = '0' then
                                O6 <= '0';
                                NEXT_STATE <= S2;
                            end if;
        when S1           => O2  <= '1';
                            if I8 = '1' or  I3 = '0' then
                                O3 <= '1';
                                NEXT_STATE <= S4;
                            end if;

        when Sn           => ...
    end case;
    wait on I0, I1, ..., Ik, CURRENT_STATE;
end process P_NEXT_STATE;
```

This modeling style also allows the output assignments to split into different
processes. This may be an interesting feature that produces a more readable
VHDL code, especially when the number of outputs is large. This may also be a
smart strategy when the nature of the outputs, combinational or sequential, is
different. In this case, outputs with the same nature may be put together in the
same process.

The next example is one in which the designer wishes to keep outputs O0
and O1 combinational, that output O2 needs to be memorized on the low level
of clock CLK, and outputs O3 and O4 have to be synchronized on the rising
edge of CLK. Three processes are written as follows

```
P_OUTPUTS_1: process
begin
    -- Default values for outputs assigned in this process
    O0  <= '0';
    O1  <= '1';
    case CURRENT_STATE is
        when S_INIT =>       O0 <= '1';
                             O1 <= '1';
        when S1           => O1 <= '1';
        when Sn           => ...
        when others  =>  null;      -- All choices (states) are not proposed.
    end case;
    wait on CURRENT_STATE;
end process P_OUTPUTS_1;

P_OUTPUTS_2: process
begin
```

```
        if CLK = '1' then
        -- Default values for outputs assigned in this process
                O2  <= '0';
            case CURRENT_STATE is
                when S_INIT  =>  O2  <= '1';
                when S5      =>  O2  <= '1';
                when Sn      =>  ...
                when others  =>  null; -- All choices (states) are not proposed.
            end case;
        end if;
        wait on CURRENT_STATE, CLK;
    end process P_OUTPUTS_2;


P_OUTPUTS_3: process
begin
        wait until CLK = '1';
        O3  <= '0';    -- Default values for outputs assigned in this process
        O4  <= '1';
        case CURRENT_STATE is
            when S_INIT      =>  O3 <= '1';    O4 <= '1';
            when S1          =>  O4 <= '1';
            when Sn          =>  ...
            when others  =>  null;        -- All choices (states) are not proposed.
        end case;
    end process P_OUTPUTS_3;
```

From the beginning of this section, finite state machines are always described by naming all possible states. For a complex control or large number of transitions, this modeling style is appropriate. Nevertheless, when the treatment is more iterative, other possibilities of description are open. Here follows a first example of a cyclic automaton: a given state always leads, conditionally or not, to the same given state. Figure 4.20 shows the description of an automaton of this category.

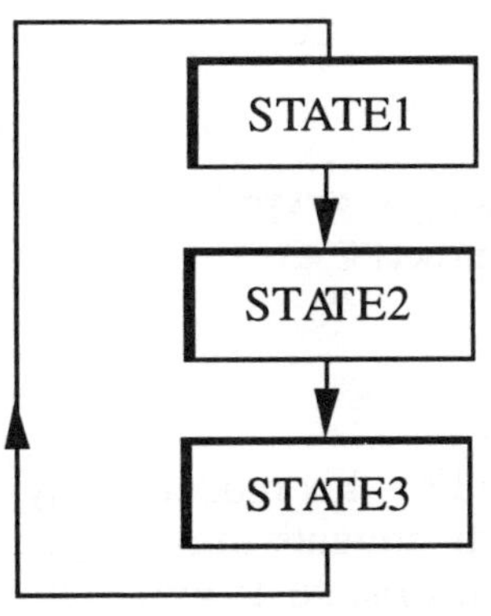


**Figure  4.20.          Simple  Cyclic  FSM**

A straightforward description of such an automaton where the states are not explicitly named (except in comments) is

```
process
begin
    wait until CLK = '1';
    -- Treatment STATE 1
    wait until CLK = '1';
    -- Treatment STATE 2
    wait until CLK = '1';
    -- Treatment STATE 3
end process;
```

Before studying more realistic examples, the question of initialization of such machines has to be solved. Two kinds of initializations are possible, synchronous or asynchronous initializations.

A synchronous initialization is used in the following example:

```
-- Signal RESET is supposed to be of type BOOLEAN.
process
begin
    loop
        wait until CLK = '1';
        if RESET then
            -- treatment of STATE 1
            exit;
        end if;
        -- treatment of STATE 2

        wait until CLK = '1';
        if RESET then
            -- treatment of STATE1
            exit;
        end if;
        -- treatment of STATE3

        wait until CLK = '1';
        loop
            -- treatment of STATE1
            exit when not RESET;
        end loop;
    end loop;
end process;
```

The redundancy implied by this modeling style is obvious. A solution may be to group the redundant treatments in a single procedure. It is only possible to apply this transformation to state 1: **exit** statements may not be put in this procedure; they have to be in the same statement part as the loop statement. The result of this transformation is

```vhdl
-- Signal RESET is supposed of type BOOLEAN
process
    procedure P_STATE1 is
    begin
            -- treatment of STATE1
    end P_STATE1;
begin
    loop
        wait until CLK = '1';
        if RESET then
            P_STATE1;
            exit;
        end if;
        -- treatment of STATE 2

        wait until CLK = '1';
        if RESET then
            P_STATE1;
            exit;
        end if;
        P_RESET;
        -- treatment of STATE 3;

        loop
            wait until CLK = '1';
            P_STATE1;
            exit when not RESET;
        end loop;
    end loop;
end process;
```

The automaton leaves STATE1 as soon as the synchronous RESET becomes
FALSE. It returns to STATE1 when the synchronous RESET becomes TRUE.

The following example presents a description of an automaton with implicit
states and asynchronous initialization.

```vhdl
process
begin

    RESET_LABEL : loop
        if RESET then
            -- Specific treatment when RESET is TRUE
        end if;

        wait until (CLK = '1' and CLK'event) or RESET;
        next RESET_LABEL when RESET;
        -- treatment of STATE 1
```

```
        wait until (CLK = '1' and CLK'event) or RESET;
        next RESET_LABEL when RESET;
        -- treatment of STATE 2

        wait until (CLK = '1' and CLK'event) or RESET;
        next RESET_LABEL when RESET;
        -- treatment of STATE 3
    end loop RESET_LABEL;

end process;
```

A realistic example is given by an automaton that performs the treatments:
- It waits for a signal RESET indicating the start of the treatment. This condition is only taken into account on a rising edge of the clock CLK (synchronous behavior). When this occurs, the internal variable ACC is initialized with the value of input I.
- During two consecutive cycles, the input data I is added to the content of variable ACC (ACC := ACC + I).
- Finally, with the next cycle, output O is set to the value ACC/2 and the automaton waits for another event on signal RESET.

The resulting state machine may be described without an explicit state variable and has a synchronous initialization on signal RESET.

```
entity ACCUMULATION is
    generic ( N   : POSITIVE;        -- Size (number of bits) of input I
              P   : POSITIVE );      -- N+P : number of bit of output O and variable ACC
    port (    CLK    : in       BIT;
              RESET  : in       BOOLEAN;
              I      : in       INTEGER range -2**(N-1) to 2**(N-1) - 1;
              O      : out      INTEGER range -2**(N+P-1) to 2**(N+P-1) - 1  );
end ACCUMULATION;

architecture A of ACCUMULATION is
begin
    process
        variable ACC : INTEGER  range -2**(N+P-1) to 2**(N+P-1) - 1;
    begin
        STATES : loop
            wait until CLK = '1';
            if RESET then
                ACC := I;
                exit STATES;
            end if;
            ACC := ACC + I;

            wait until CLK = '1';
            if RESET then
                ACC := I;
```

```
                    exit STATES;
                end if;
                ACC := ACC + I;

                wait until CLK = '1';
                if RESET then
                    ACC := I;
                    exit STATES;
                end if;
                O := ACC / 2;

                STATE_1 : loop
                    wait until CLK = '1';
                    ACC := I;
                    exit STATE_1 when not RESET;
                end loop STATE_1;
            end loop STATES;
        end process;
    end A;
```

Although rendered more complex by synchronous initialization, such a description is relatively simple to read and easy to debug and maintain.

A similar example involving an asynchronous initialization may be described as follows:

```
architecture B of ACCUMULATION is
begin
    process
        variable ACC : INTEGER  range -2**(N+P-1) to 2**(N+P-1) - 1;
    begin
        RESET_LABEL : loop
            -- No specific treatment for  RESET = TRUE
            wait until (CLK = '1' and CLK'event) or RESET;
            next RESET_LABEL when RESET;
            ACC := I;
            wait until (CLK = '1' and CLK'event) or RESET;
            next RESET_LABEL when RESET;
            ACC := ACC + I;
            wait until (CLK = '1' and CLK'event) or RESET;
            next RESET_LABEL when RESET;
            ACC := ACC + I;
            wait until CLK = '1' and CLK'event or RESET;
            next RESET_LABEL when RESET;
            O <= ACC / 2;
        end loop RESET_LABEL;
    end process;
end B;
```

Another example points out the flexibility and richness of the possible styles. The use of **loop** statements and **if** or **case** statements allows for keeping very close to the original algorithm. This is useful in the signal processing domain. A usual functionality in signal processing is to compute a convolution. From a mathematical point of view, a convolution may be expressed as

$$C = \sum a_i * x_i$$

with i varying from 0 to N-1, where $x_i$ are the input data sampled with clock CLK, and $a_i$ are fixed (or varying) coefficients.

```vhdl
package P is
        type TYPE_A_INTEGER is array (NATURAL range <>) of INTEGER;
end P;

use WORK.P.all;
entity CONVOLUTION is
        generic ( A C : TYPE_A_INTEGER ;        -- Coefficient array
                  N    : POSITIVE;              -- Size (number of bits) of input I
                  P    : POSITIVE );            -- N+P : number of bits of output O and ACC
        port (    CLK      : in       BIT;
                  RESET  : in       BOOLEAN;
                  X        : in       INTEGER range -2**(N-1) to 2**(N-1) - 1;
                  C        : out      INTEGER range -2**(N+P-1) to 2**(N+P-1) - 1  );
end CONVOLUTION;

architecture A of CONVOLUTION is
begin
        process
        begin
                loop
                    wait until CLK = '1';
                    C <= X * AC(0);
                    exit when  not RESET;
                end loop;

                for I in 1 to AC'high loop
                    wait until CLK = '1';
                    if RESET then
                        C <= X * AC(0);
                        exit;
                    end if;
                    C <= C + X * AC(I);
                end loop;
        end process;
end A;
```

In this description, the most expensive inferred hardware operator is the multiplication operator. This operator may have to be shared: it is only used once during each period.

This modeling style seems to make resource sharing easier. Indeed, such a description is close to a purely concurrent one. By deleting some synchronization statements, the description becomes fully parallel and therefore combinational.

State machines may therefore be described in two ways:
- by explicitly defining a state variable (of a scalar and therefore enumeratable type) and updating it synchronously depending on the input values and the state variable current value.
- by writing an algorithm with multiple synchronization points, each of these points representing a state of the system.

In the general case, the first solution is simpler to write. In the second option redundancy is introduced when expressing the synchronous or asynchronous initializations.

As previously seen, writing large automatons, with many states and initialization conditions, may be complex. Factorization may be a way of simplifying the writing:

- For example, when a set of treatments is common to several states as follows

```
    ...
        when STATE_i =>    statement_1;
                           statement_2;
                           statement_i;
    ...
        when STATE_k =>    statement_1;
                           statement_2;
                           statement_k;
    ...
```

It is interesting to put together the common treatment in a single procedure

```
procedure P_STATEMENTS_12 is
begin
        statement_1;
        statement_2;
end P_STATEMENTS_12;
    ...
        when STATE_i =>    P_STATEMENTS_12 ;
                           statement_i;
    ...
        when STATE_k =>    P_STATEMENTS_12 ;
```

```
        statement_k;
    ...
```

- It may appear that a sequence of states has to be reused. Factorization is also possible in this case.

  This concept is close to the notion of subprograms in computers. The program counter of a computer may be seen as the state register of a finite state machine. Its content may be stored in a specific memory: a stack. This allows the computer to execute the sub-sequence. Once this execution is completed, the stored value of the program counter is restored by the stack. This concept is used in the following lines:

```
    ...
    case STATE is
        when STATE_1  =>  STATE := STATE_2;
        when STATE_2  =>  STACK := STATE_3;
                          STATE := SUB_STATE_1;
        when STATE_3  =>  STATE := STATE_4;
        ...
        when STATE_i  =>  STACK := STATE_k;
                          STATE := SUB_STATE_1;
        when STATE_k  =>  STATE := STATE_n;
        when STATE_n  =>  ...

        when SUB_STATE_1 =>  STATE := SUB_STATE_2;
        when SUB_STATE_2 =>  ...
        ...
        when SUB_STATE_n =>  STATE := STACK;
        ...
    end case;
```

The subsequence of states SUB_STATE_1 to SUB_STATE_n is executed after states STATE_2 or STATE_i. Once the subsequence in state SUB_STATE_n, the state variable STATE is updated with the value of the state to be executed when returned. The value of this state is at the top of STACK.

This modeling style is well suited to describing a complex sequencer in which subsequences are often the same. It has to be noted that no parallelism (or concurrency) occurs in such a description: the automaton is always in a unique state at a given time. If the need to use a subsequence in parallel is real, the complete modeling style has to be changed and the notion of hierarchical automatons introduced.

In this context, a master automaton controls the activity of a slave automaton. Different control protocols are possible, from handshaking to synchronous or asynchronous forcings. Explicit control signals are used for this purpose. Of course, a master automaton may perfectly control several slave automatons.

# 5. DESIGN METHODOLOGY

Knowledge of the language, its syntax subset accepted by synthesis tools as well as the main possible modeling styles to infer a given hardware have been introduced throughout the first chapters of this book.

It is now time to look at how to integrate the synthesis process within an existing design methodology. The key questions may be summarized as follows:
- What are the steps of such a new methodology?
- What can really be expected from commercial synthesis tools?
- What are their main characteristics?

## 5.1. SYNTHESIS DESIGN CYCLE

### 5.1.1. Design Cycle Steps

When performed well, these steps lead to an efficient design. They may be divided into design steps and validation steps. Validation steps are essential to ensure that the inferred hardware has the desirable properties whatever the quality of the synthesis tools. Indeed, differences may appear between the before-synthesis and after-synthesis behaviors.

Figure 5.1 illustrates the design cycle. Such a cycle may be used for ASIC design as well as for programmable circuit (FPGA, EPLD) design.

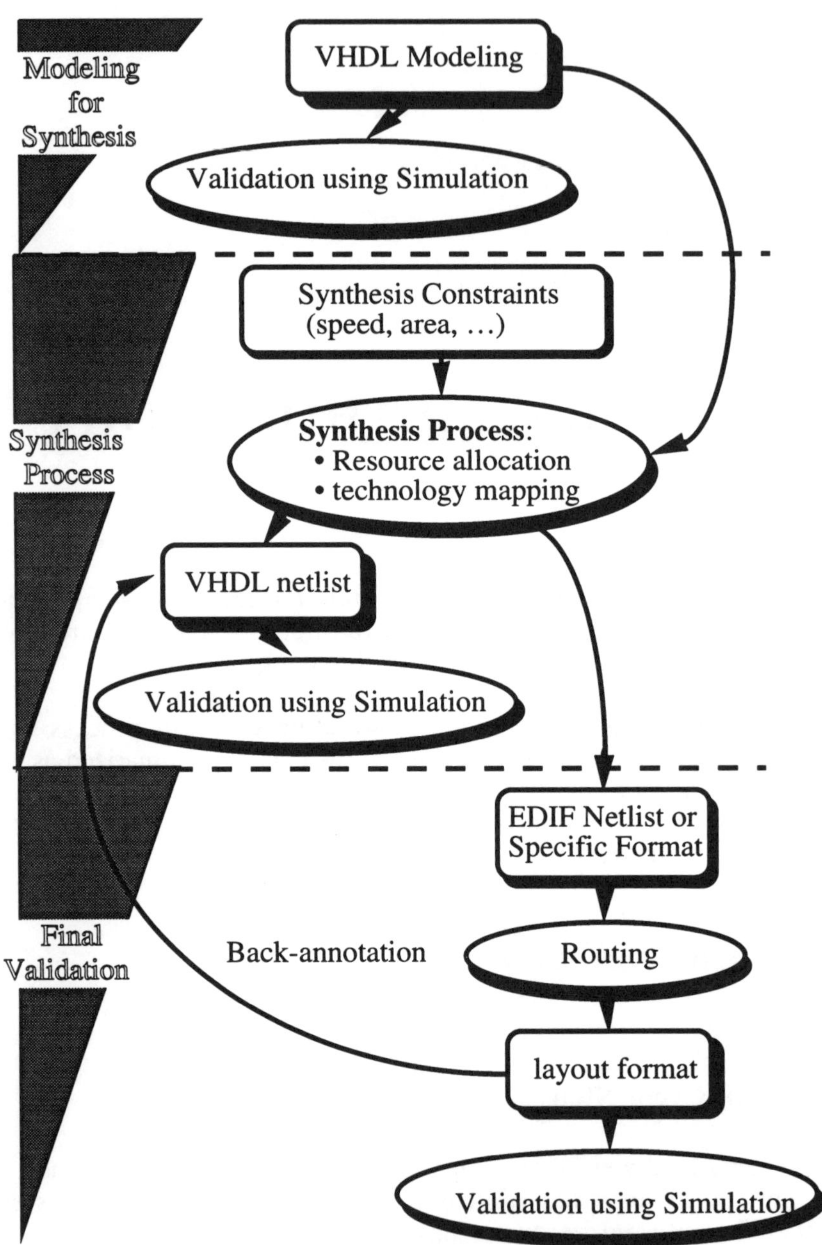


**Figure 5.1. Design Cycle Flow Chart**

Three phases may be distinguished in this design cycle: modeling for synthesis, synthesis process, and final validation.

## 5.1.2.  Modeling for Synthesis

Modeling for synthesis is the first phase. At this level, the specifications have to be clear and are coded in VHDL. The modeling style used is important, and this modeling task consists mainly in finding a good tradeoff between the abstraction level and the accuracy of the description.

The choice of data types (integers, enumerated types, composite types, and so on) and the choice of VHDL constructs (subprograms, loop statements, and so on) are essential.

The most abstract is the description, and the easiest are its validation, its maintainability, and its understanding by somebody else. A more accurate description involves more details. These details usually have a negative effect on the readability and maintainability of the description but may significantly increase the control of the inferred hardware. Very often, several loops back between the modeling phase and the analysis of the synthesis result simulation are necessary to make a given part of the original description more accurate (and thus less abstract) to obtain an optimal hardware counterpart.

The model finally obtained is validated using simulation. For this task, all the power of VHDL to describe the specific simulation environment of the application (testbench) is available to the designer. VHDL has the great advantage of providing such possibilities: in certain domains, the modeling of the testbench (stimulis generation, trace analysis) may be complex and involve more lines of code than the model itself.

## 5.1.3.  Synthesis Process

The second phase of the synthesis design cycle is the synthesis process itself. The VHDL code, validated during the previous phase, is translated into a generic netlist. Memorization elements as well as operative resources appear explicitly. The whole behavior is now described as calls to elements of the generic library. The richness of this library is directly related to the knowhow of the tool.

This concept of generic netlist allows for other transformations depending on constraints given by the designer. For example, an addition is implemented using a "Ripple Carry" adder if the speed constraints are easy enough to meet. A "Carry Look Ahead" adder is used in the opposite case.

Specific optimizations are performed on Boolean expressions. Their goals are multiple. When targeted to reduce the quantity of inferred hardware, the entire set of Boolean equations is minimized by grouping common terms and simplifying. When the idea is to achieve a better control of the timing performances of the inferred hardware, the number of logic gate levels is reduced even if this transformation implies duplicating certain expressions (and thus increasing the inferred hardware).

Finally, the resulting description has to be mapped on the physical target library. This step is called "technology mapping." The result of the synthesis process has to fit the area and speed constraints. Technology mapping algorithms are usually timing driven and therefore give a higher priority to the timing constraints over the area constraints: the timing constraints have to be strictly respected with the smallest area possible.

At this step, synthesis tools have the capability to propose a first estimation of the cost (in terms of area) of the route-and-place operation. This capability allows for more realistic results.

The result of the synthesis process is a netlist, which is processed by place-and-route tools. The EDIF format, a standard format originally created for schematic description exchanges, is generally used to code this netlist.

Indeed, this netlist may also be used to validate the synthesis result. This validation is usually performed using simulation. Two different approaches are possible for this simulation: a new simulation in the specific context of the CAD tool or a simulation with the VHDL testbench used on the description before synthesis. This second approach has to take priority, since it avoids re-writing the stimulis. This approach implies that the synthesis tool is able to generate a VHDL netlist and proposes VHDL descriptions of the physical library elements. These VHDL descriptions have to provide the behavior in time of these elements (timing characteristics of the physical library elements).

"After synthesis" simulation allows for checking that the right behavior is obtained but also provides a first estimation of the behavior in time. What are the possible differences between the results of a simulation before synthesis and those of a simulation of the synthesis result, after technology mapping?

Different problems may occur depending on the logic type chosen for synthesis:

- If type BIT is used, very few differences are to be expected. If before-synthesis and after-synthesis behaviors are not consistent, the frequency of the clock, possibly too high, has to be checked. Drastically reducing the clock frequency to check the function is an efficient way to exclude (or not) a clock problem. Unfortunately, if inconsistencies are still encountered, no good strategy exists to systematically isolate the source of the problem. Statistically, a large number of the remaining errors at this step of the design are related to a bad initialization of a memorization element. Such an element has often been described before synthesis with an abstract modeling style and its initialization has been performed by default, without explicit statement. In order to help to detect non-initializations, the model may be rewritten using types STD_LOGIC or STD_ULOGIC. Any noninitialization in the resulting description will appear as a propagation of value 'U' on all dependent signals. Returning back over track to the error will be easier.

- If types STD_LOGIC or STD_ULOGIC have been initially chosen for modeling, noninitialization problems are therefore easily identified. Indeed, such a problem may be linked to the simulation itself and the behavior perfectly correct in real applications. To be sure of the consistency between the descriptions before and after synthesis, it will nevertheless be essential to force the initial value of the memorization element in question. This forcing is performed in the description before synthesis. This task may appear cumbersome but it is the price to pay to ensure the quality of the synthesized design.

### 5.1.4.  Final Validation

This validation takes place after the place-and-route operation has been performed. Its goal is to check that the original functionality has been respected and that the timing characteristics of the generated hardware meet the timing constraints. Two methods of achieving this goal are possible:

- performing the simulation within the proprietary environment. In this case, the simulator uses the values of the net loads computed after the place-and-route operation.
- back annotating the VHDL netlist resulting from the synthesis process with the characteristics deduced from the place-and-route operation. The great advantage of this second method is to be able to remain in the VHDL simulation framework.

It is interesting to note that information provided by place-and-route tools may be useful for synthesis tools. For example, more accurate information on the load of a given port may allow an extra optimization. To avoid synthesis-routing cycles, which may possibly not converge, a new approach is now proposed by the latest synthesis tools. Their strategy mainly consists in heavily coupling synthesis and place-and-route tools. Place-and-route evaluations are available throughout the synthesis process and are taken into account during synthesis tool optimizations. Finally, the synthesis tool not only generates a netlist but also a set of place-and-route constraints in order to optimize such and such a critical path.

## 5.2. SYNTHESIS PROCESS CONTROL

### 5.2.1. What Is Synthesis Process Control?

The VHDL description, the model, is the essential input of synthesis tools: it includes a description of the desired behavior. Since a hardware device has

also to respect other constraints, a means of expressing them is required. Indeed, no standard language to do this is available and each synthesis platform offers its own proprietary solution. Whatever the solution, two kinds of constraints are usually encountered:
- Constraints on physical characteristics of the design (timing, area, loads).
- Constraints (options) on the design flow itself (interface formats, control of the hierarchy).

## 5.2.2. Design   Constraints

When no constraint at all is given to the synthesis process, synthesis tools minimize the cost of the generated hardware. There are different ways to compute (or evaluate) the cost of generated hardware. When designing an ASIC, this cost is simply related to area and either expressed in square-microns or in number of basic gates (minimal size inverters or two-input nand gates for example). If a programmable component (FPGA or EPLD) is targeted, the cost evaluation is more complex since its value is not continuous: the targets are a range of standard circuits, and one of them has to be chosen. The number of basic elements is not the only criteria for this choice: a place-and-route evaluation has to exist.

There are many possible physical constraints. Some of them allow for description of the environment of the targeted hardware in real-life, others indicate performances to be attained.

The first category includes the following:
- The value of capacitive load on input or output ports
- The value of input or output port resistances
- The temperature range for "real-life" use
- The supply voltage

In the second category, the following indications may be found:
- Min or max pin-to-pin timing between specific inputs/outputs
- Min or max propagation delay on a given wire
- Clock frequency
- Value of the acceptable clock skew, i.e., the acceptable phasing between all wires carrying the same clock everywhere in the circuit. This constraint has to be taken into account in the place-and-route strategy and when sizing the amplifiers.

## 5.2.3. Interface   Formats

Different formats (and languages) may be encountered when describing the synthesis process inputs.

- A behavioral model is usually written in VHDL or Verilog. Since this book is devoted to VHDL modeling, readers interested in entering Verilog descriptions may refer to the comparison VHDL/Verilog proposed in chapter 9 of [BER92].[8]
- If the goal of the synthesis process is to translate an already existing design into another format (this is a possible use of the synthesis process), the input format may be EDIF, a PLA format or a VHDL structural description.
- Among these input formats is the description of the physical library (containing characterized elements). Unfortunately, there is no standard format for this purpose and proprietary ones are used.
- No standardization exists in the expression of the design constraints themselves. They may appear as user-defined VHDL attributes, proprietary commands, or syntactic comments.

The following interface formats are used for synthesis process outputs:
- The netlist, result of the synthesis process, is usually provided in EDIF for place-and-route tools and in VHDL (or Verilog) for after-synthesis simulation purposes.
- The place-and-route constraints have no standard format. Proprietary descriptions are generally proposed.
- Rough estimations of place-and-route characteristics such as net and cell delays, net resistances, net loads, or timings are stored using SDF standard format.

---

[8] J. M. Bergé, A. Fonkoua, S. Maginot, J. Rouillard, "VHDL Designer's Reference," Kluwer Academic Publishers, June 1992.

# 6. SYNTHESIS STANDARD ENVIRONMENT

## 6.1. PRINCIPLE

This environment is not part of the language and is therefore not defined within the LRM. It appears as separate standard packages that have been (or will soon be) balloted and are (will be) available on any VHDL synthesis platform.

This environment consists of
- A multivalue logic system adapted to synthesis of common digital devices, the 9-state STD_LOGIC_1164 package. Already accepted as an IEEE standard, this package is proposed by all simulation and synthesis platforms. The first part of this chapter presents this package.
- The results of the Synthesis Special Interest Group (SSIG) of IEEE. The work of this group covers different axes:
  - A document defining the synthesis semantics of the types defined in the STD_LOGIC_1164.
  - A set of synthesizable primitives working on numeric types: the standard arithmetic packages.
  - Specific user-defined attributes and their semantics. They are a standard way of giving specific information to the synthesis tool.

  The work of this working group is just about to be submitted to the balloting vote (Autumn 1993), and it is therefore too early to detail its contents. Nevertheless, since it appears that this initiative receives much support from designers and tool vendors, it is interesting to present the spirit and trends of this work. This constitutes the second part of this chapter.

## 6.2. PACKAGE  STD_LOGIC_1164

Synthesis has not been the only aim when defining this package and this fact explains the lack of a precise synthesis semantics. A standard multivalue logic system offers many other advantages such as facilitating the exchange of VHDL descriptions or allowing tool builders to propose more efficient tools based on optimizing operations on the proposed logic type.

### 6.2.1. Logic  Type  Interpretation:  Simulation  Semantics

The whole package is based on an enumerated type called STD_LOGIC (or STD_ULOGIC when unresolved), which has nine values. This section only deals with the simulation semantics of this type:

```
type STD_ULOGIC is (   'U',   -- Uninitialized
                       'X',   -- Forcing Unknown
                       '0',   -- Forcing 0
                       '1',   -- Forcing 1
                       'Z',   -- High Impedance
                       'W',   -- Weak Unknown
                       'L',   -- Weak 0
                       'H',   -- Weak 1
                       '-'    -- Don't care        ) ;
```

The 'U' state, as the leftmost value of STD_LOGIC, is the default value of any object declared of this type. Called "uninitialized" state, its semantics is simple: if, during the simulation, such a value appears, the corresponding object has not been initialized, thus has never been assigned.

The semantics of the 'X' state, the "Unknown" state, is very close to the previous one. It does not imply an electrical value but the lack of information on its value. Indeed, the only information provided by the value 'X' is that this object has been initialized.

Values '0', '1' and 'X' have "weak values" associated: respectively 'L', 'H' and 'W'. The purpose of this duplication is to allow more accurate simulation taking into account a first notion of strength. All their semantics is contained within the resolution table given in figure 6.1.

### Remark:

'X', 'W' and 'U' are called metalogical values, and their nature is different from the other ones. The designer has to be very careful when using them. For example, a simple comparison operator may lead to a trap: 'U' < '0' is always true (with its simulation semantics) due to the order of the enumerated type, but it has no real meaning.

The value '-', a "don't care" value, has no real simulation meaning. It purpose is to leave the synthesis tool free to perform optimizations by replacing it with '0' or '1'. This synthesis semantics is studied in paragraph 3.2.3.

```
    -   U   X   0   1   Z   W   L   H   -       -       -
    ------------------------------------------------------------------
    (  'U', 'U', 'U', 'U', 'U', 'U', 'U', 'U', 'U'  ),  --  U   -
    (  'U', 'X', 'X', 'X', 'X', 'X', 'X', 'X', 'X'  ),  --  X   -
    (  'U', 'X', '0', 'X', '0', '0', '0', '0', 'X'  ),  --  0   -
    (  'U', 'X', 'X', '1', '1', '1', '1', '1', 'X'  ),  --  1   -
    (  'U', 'X', '0', '1', 'Z', 'W', 'L', 'H', 'X'  ),  --  Z   -
    (  'U', 'X', '0', '1', 'W', 'W', 'W', 'W', 'X'  ),  --  W   -
    (  'U', 'X', '0', '1', 'L', 'W', 'L', 'W', 'X'  ),  --  L   -
    (  'U', 'X', '0', '1', 'H', 'W', 'W', 'H', 'X'  ),  --  H   -
    (  'U', 'X', 'X', 'X', 'X', 'X', 'X', 'X', 'X'  )   --  -   -
```

**Figure 6.1. Resolution Function of type STD_LOGIC**

## 6.2.2. Description of Package STD_LOGIC_1164

This package does not only export the nine-state logic types in their resolved and unresolved forms (and the corresponding resolution function). Associated bit arrays, logical operators and conversion functions are also provided. Its package declaration is

```
package STD_LOGIC_1164 is
---------- Logic state system (unresolved) ------------------------------------------------------
type STD_ULOGIC is (  'U',   -- Uninitialized
                      'X',   -- Forcing Unknown
                      '0',   -- Forcing 0
                      '1',   -- Forcing 1
                      'Z',   -- High Impedance
                      'W',   -- Weak Unknown
                      'L',   -- Weak 0
                      'H',   -- Weak 1
                      '-'    -- Don't care         ) ;
---------- Unconstrained array of STD_ULOGIC for use with the resolution function ----------------------
type STD_ULOGIC_VECTOR is array (NATURAL range <>) of STD_ULOGIC ;
---------- Resolution function ------------------------------------------------------------------
function RESOLVED( S : STD_ULOGIC_VECTOR) return STD_ULOGIC ;
---------- *** Industry standard logic type *** -------------------------------------------------
subtype STD_LOGIC is RESOLVED STD_ULOGIC ;
---------- Unconstrained array of STD_LOGIC for use in declaring signal arrays ---------------------
```

```
type STD_LOGIC_VECTOR is array (NATURAL range <>) of STD_LOGIC ;
---------- Common subtypes ----------------------------------------------------
subtype X01     is RESOLVED STD_ULOGIC range 'X' to '1' ; -- ('X', '0', '1')
subtype X01Z    is RESOLVED STD_ULOGIC range 'X' to 'Z' ; -- ('X', '0', '1', 'Z')
subtype UX01    is RESOLVED STD_ULOGIC range 'U' to '1' ; -- ('U', 'X', '0', '1')
subtype UX01Z   is RESOLVED STD_ULOGIC range 'U' to 'Z' ; -- ('U', 'X', '0', '1', 'Z')
---------- Overloaded logical operators ----------------------------------------
function "and"  ( L : STD_ULOGIC ; R : STD_ULOGIC )     return   UX01 ;
function "nand" ( L : STD_ULOGIC ; R : STD_ULOGIC )     return   UX01 ;
function "or"   ( L : STD_ULOGIC ; R : STD_ULOGIC )     return   UX01 ;
function "nor"  ( L : STD_ULOGIC ; R : STD_ULOGIC )     return   UX01 ;
function "xor"  ( L : STD_ULOGIC ; R : STD_ULOGIC )     return   UX01 ;
function "not"  ( L : STD_ULOGIC                  )     return   UX01 ;
---------- Vectorized overloaded logical operators -----------------------------
function "and"  ( L, R : STD_LOGIC_VECTOR  )     return   STD_LOGIC_VECTOR ;
function "and"  ( L, R : STD_ULOGIC_VECTOR )     return   STD_ULOGIC_VECTOR ;
function "nand" ( L, R : STD_LOGIC_VECTOR  )     return   STD_LOGIC_VECTOR ;
function "nand" ( L, R : STD_ULOGIC_VECTOR )     return   STD_ULOGIC_VECTOR ;
function "or"   ( L, R : STD_LOGIC_VECTOR  )     return   STD_LOGIC_VECTOR ;
function "or"   ( L, R : STD_ULOGIC_VECTOR )     return   STD_ULOGIC_VECTOR ;
function "nor"  ( L, R : STD_LOGIC_VECTOR  )     return   STD_LOGIC_VECTOR ;
function "nor"  ( L, R : STD_ULOGIC_VECTOR )     return   STD_ULOGIC_VECTOR ;
function "xor"  ( L, R : STD_LOGIC_VECTOR  )     return   STD_LOGIC_VECTOR ;
function "xor"  ( L, R : STD_ULOGIC_VECTOR )     return   STD_ULOGIC_VECTOR ;
function "not"  ( L    : STD_LOGIC_VECTOR  )     return   STD_LOGIC_VECTOR ;
function "not"  ( L    : STD_ULOGIC_VECTOR )     return   STD_ULOGIC_VECTOR ;
---------- Conversion functions ------------------------------------------------
function TO_BIT ( S : STD_ULOGIC ; XMAP : BIT :='0' ) return   BIT ;
function TO_BITVECTOR    ( S : STD_LOGIC_VECTOR ; XMAP : BIT :='0' )
                                                return   BIT_VECTOR ;
function TO_BITVECTOR    ( S : STD_ULOGIC_VECTOR ; XMAP : BIT :='0' )
                                                return BIT_VECTOR ;
function TO_STDULOGIC     ( B : BIT )            return   STD_ULOGIC ;
function TO_STDLOGICVECTOR ( B : BIT_VECTOR ) return   STD_LOGIC_VECTOR ;
function TO_STDLOGICVECTOR ( S : STD_ULOGIC_VECTOR )
                                                return   STD_LOGIC_VECTOR ;
function TO_STDULOGICVECTOR (B : BIT_VECTOR) return   STD_ULOGIC_VECTOR ;
function TO_STDULOGICVECTOR ( S : STD_LOGIC_VECTOR )
                                                return   STD_ULOGIC_VECTOR ;
---------- Strength strippers and type converters ------------------------------
function TO_X01    ( S : STD_LOGIC_VECTOR )     return   STD_LOGIC_VECTOR ;
function TO_X01    ( S : STD_ULOGIC_VECTOR )    return   STD_ULOGIC_VECTOR ;
function TO_X01    ( S : STD_ULOGIC )           return   X01 ;
function TO_X01    ( B : BIT_VECTOR )           return   STD_LOGIC_VECTOR ;
function TO_X01    ( B : BIT_VECTOR )           return   STD_ULOGIC_VECTOR ;
function TO_X01    ( B : BIT )                  return   X01 ;

function TO_X01Z   ( S : STD_LOGIC_VECTOR )     return   STD_LOGIC_VECTOR ;
function TO_X01Z   ( S : STD_ULOGIC_VECTOR )    return   STD_ULOGIC_VECTOR ;
function TO_X01Z   ( S : STD_ULOGIC )           return   X01Z ;
```

```
function TO_X01Z    ( B : BIT_VECTOR )            return    STD_LOGIC_VECTOR ;
function TO_X01Z    ( B : BIT_VECTOR )            return    STD_ULOGIC_VECTOR ;
function TO_X01Z    ( B : BIT )                   return    X01Z ;
-------------------------------------------------------------------------------------
function TO_UX01    ( S : STD_LOGIC_VECTOR )   return    STD_LOGIC_VECTOR ;
function TO_UX01    ( S : STD_ULOGIC_VECTOR ) return    STD_ULOGIC_VECTOR ;
function TO_UX01    ( S : STD_ULOGIC )        return    UX01 ;
function TO_UX01    ( B : BIT_VECTOR )        return    STD_LOGIC_VECTOR ;
function TO_UX01    ( B : BIT_VECTOR )        return    STD_ULOGIC_VECTOR ;
function TO_UX01    ( B : BIT )               return    UX01 ;
---------- Edge detection --------------------------------------------------------
function RISING_EDGE   ( signal S : STD_ULOGIC ) return    BOOLEAN ;
function FALLING_EDGE ( signal S : STD_ULOGIC ) return    BOOLEAN ;
---------- Object contains an unknown --------------------------------------------
function IS_X    ( S : STD_ULOGIC_VECTOR )      return    BOOLEAN ;
function IS_X    ( S : STD_LOGIC_VECTOR )       return    BOOLEAN ;
function IS_X    ( S : STD_ULOGIC )             return    BOOLEAN ;
-------------------------------------------------------------------------------------

end STD_LOGIC_1164 ;
```

## 6.3. SYNTHESIS WORKING GROUP RESULTS

Since this work is in progress. The information contained in this section may change. The only purpose here is to make the designer aware of what is being discussed and decided in this IEEE working group to help him understand the future trends in synthesis standard environment.

### 6.3.1. Logic Type Interpretation: Synthesis Semantics

The purpose of the synthesis group effort on logic type interpretation is not only to give a semantics to each literal of the enumerated type STD_LOGIC but also to propose a single interpretation of VHDL features where they are involved.

Values '0' and '1' have to be respectively interpreted by synthesis tools as "tied to ground" and "tied to VCC." No specific semantics is provided for values 'L' and 'H'. Synthesis tools are free to interpret them. A possible strategy for tool builders is to consider 'L' as '0' and 'H' as '1', but this has to be checked by the designer.

**Remark:**
Initial and default values are real issues in synthesis. Certain designers use them to model an "initial reset" of their system but no consensus has been reached on this practice. Therefore, no standard semantics is associated with them. Designers have to be aware that any description using such initialization is not portable in terms of synthesis results. Indeed, it is not a good practice to use initial values in a description if the functionality of the description depends on them. If really needed, initial values have to be carefully used: they imply different behavior in the first simulation steps of the input and output descriptions of the synthesis process.

For the other STD_ULOGIC values, 'Z', 'U', 'W', and 'X' (the value '-' will be studied further), the semantics depends on their use. Different cases may be distinguished:

- Such a value is used in a default expression or as initial value. The remark above applies to this case. No semantics is defined.
- Such a value is assigned to a variable or a signal. In that case, values 'U', 'W', and 'X' (also called metalogical values) are treated as if they were "don't care" values. The "don't care" value ('-'), may be replaced by the synthesis tool by either '0' or '1' (see below). Value 'Z' implies the synthesis tool to infer a three-state buffer. The output of this buffer is the target of the assignment.
- Such a value is used for implicit comparison purposes in **case** statements. The statements were they are involved are simply ignored.
- Such a value is used with equality or comparison operators (this involves **if** statements). The expression A=B, respectively expression A/=B, returns FALSE, respectively TRUE, if one of the operands has such a value ('U', 'W' or 'X'). Comparison with operators ">", ">=", "<", "<=" systematically returns FALSE.

The "don't care" value ('-') is useful in the synthesis domain. Indeed, it provides different functionalities:

- When occurring as the right part of an assignment, optimization is allowed.

```
S <= '-';
V := '-';
ARRAY_S <= "----";
```

The synthesis tool is free to replace each "don't care" character by either '0' or '1'. The purpose of this flexibility is to allow optimizations and to save hardware resources. It has to be noted that when a "don't care" value is "replaced" by '0' (or '1'), no new change is possible.

- When compared to other values, it is a wild card.
  This may happen implicitly in a **case** statement or explicitly in an **if** statement or in any expression involving comparison operators. The purpose of wild cards is to filter some values. For example, "1101" and "1--1" match because the comparison is filtered; only the first and the last bits are compared. "1101" and "10-1" values do not match because of the second bit.

It is easy to understand the interest of such a wild card functionality for synthesis purpose. Unfortunately, this introduces an important difference between simulation and synthesis semantics.

When writing the expression "1101"="1--1", its evaluation during simulation concludes to FALSE as synthesis tools evaluate it as TRUE. Obviously, this situation may lead to different behavior of the description, before and after synthesis process.

To avoid this, the use of a specific function called STD_MATCH is recommended by the IEEE working group. This function is provided by a standard package called STD_SYNTHESIS_1164.[9] Its purpose is to define the simulation semantics of a wild card to the "don't care" ('-') value. Using this function when comparing to "don't care" values ensures the consistency between simulation and synthesis semantics. In the following example, the synthesis result is the same whatever line is written (line 1 or line 2) but only line 2 ensures consistency between simulation and synthesis.

```
signal A: STD_ULOGIC_VECTOR (3 downto 0);
...
...
if      A /= "10--" then                      -- Line 1
...
...
if      STD_MATCH(A, "10--") then             -- Line 2
...
```

The use of function STD_MATCH is therefore strongly recommended. Package STD_SYNTHESIS_1164 proposes different overloadings of the STD_MATCH function: the input parameters are either of types STD_ULOGIC or STD_ULOGIC_VECTOR (as used in the previous example) and the return value is either a boolean (as also used in the previous example) or the value '0' or '1' of type STD_ULOGIC.

---

[9] To be standardized (Autumn 1993).

## 6.3.2. Arithmetic  Packages

### 6.3.2.1. Contents  of  these  packages

The arithmetic environment is composed of two packages, which provide up to 250 subprograms. Both package declarations and bodies are standard but synthesis tool vendors are free to use their own bodies for optimization reasons. Of course, in this case, the behavior of the subprogram has to be consistent with those defined within the standard VHDL package body. These packages export arithmetic operations (and some conversion functions) on vectors of bits.

The two packages, NUMERIC_BIT and NUMERIC_STD, have the same constitution. Indeed, their main difference is that package NUMERIC_BIT deals with vectors whose elements are of type BIT and package NUMERIC_STD performs operations on vectors of type STD_ULOGIC. They may be considered as symmetric packages.

When using package NUMERIC_BIT, the designer imports arithmetic subprograms dealing with BIT_VECTOR and also with two bit-array types: SIGNED and UNSIGNED. The first lines of the package NUMERIC_BIT declaration are

```
package NUMERIC_BIT is
        type UNSIGNED is array (NATURAL range <>) of BIT;
        type SIGNED is array (NATURAL range <>) of BIT;

        ...
```

Follow the exported arithmetic operations, such as +, -, *, /, >, >=, <, >=, =, abs, rem, mod, and many conversion functions between types. The large number of exported subprograms is easily understandable

- These operations are both defined as functions (operator overloading) and as procedures (to return extra information such as the length of the result).
- These operations are defined for operands of type BIT_VECTOR (indeed, only addition and subtraction are defined for type BIT_VECTOR), SIGNED, UNSIGNED, and INTEGER. Each subprogram is therefore overloaded tens of times.

Symmetrically, package NUMERIC_STD deals with bit-array type STD_ULOGIC_VECTOR, defined in package STD_LOGIC_1164, and also with two other bit-array types: SIGNED and UNSIGNED. This time, the elements of these types are of type STD_ULOGIC. The first lines of package NUMERIC_STD are therefore

```
library IEEE;
use IEEE.STD_LOGIC_1164.all;
package NUMERIC_STD is
        type UNSIGNED is array (NATURAL range <>) of STD_ULOGIC;
        type SIGNED is array (NATURAL range <>) of STD_ULOGIC;
```

**Remark:**
It may appear strange to base the array types UNSIGNED and SIGNED on elements of type STD_ULOGIC, which is unresolved. Indeed, a big discussion took place within the synthesis working group on this topic. Is the resolved type STD_LOGIC more convenient? The main reason to promote an unresolved type is related to the safety of the resulting descriptions. If the designer makes an error in a netlist by connecting two outputs together, the use of an unresolved array type will automatically generate a compilation error. If elements were resolved, such an error would be detected later in the design cycle and would probably imply the use of debugging tools to localize it. The price to pay for this safety feature is a cumbersome type conversion when encountering three-state mechanisms.

### 6.3.2.2. Conventions on numeric types

It has been decided that the most significant bit (MSB) position corresponds to the 'LEFT element of the array-type.

Concerning negative numbers, the convention for their encoding is the two's complement.

### 6.3.2.3. Convention on numeric operations

Global strategies are defined for all numeric operations:
- Since it is not obvious to define how to fill or extend a bit-array to fit a given length, it has been decided that numeric operations will use different operand size when necessary.
- When returning a result, arithmetic subprograms may only use elements of the "computed value subset": '0', '1', 'X', 'U'.
- Dealing with bit-arrays containing 'X' elements (unknown) requires a clear propagation strategy for this value during simulation. As shown in Figure 6.2, at least three reasonable strategies were possible. Finally, the "least pessimistic" strategy has been chosen. This solution provides the most accurate result.

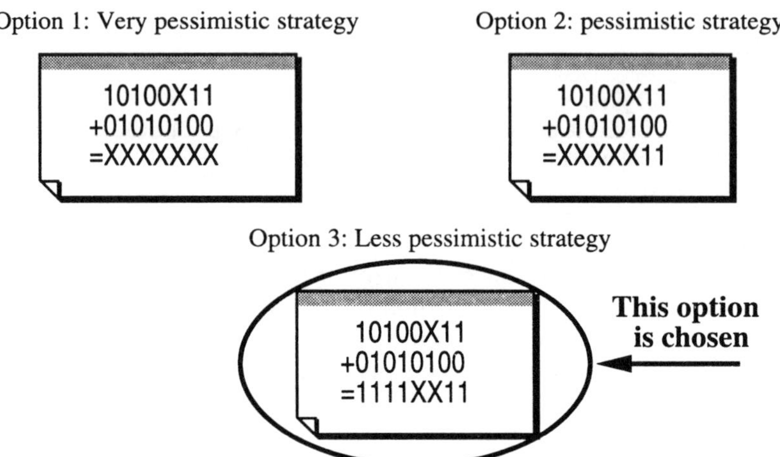



**Figure 6.2. 'X' Propagation Strategy, Example of the Addition**

## 6.3.3. Special Identifications

At the beginning of the SSIG work, the definition of many user-defined attributes was envisaged to closely control the synthesis process. The goal of these attributes was to ensure a better interoperability between synthesis tools. Among the proposed attributes were attributes to control the hierarchy, to recognize a clock or a reset, to specify how to encode a given state, etc.

Indeed, finding a common subset of attributes to suit everybody soon proved very difficult. Too many tools and sometimes underlying design methodologies were used. For example, certain designers use special identifications (by attributes or comments) to specify clocks whereas others simply rely on their synthesis tools.

Finally, no standard will be provided for standard identifications.

# 7. CASE STUDY

## 7.1. TRAFFIC LIGHT CONTROLLER: ONCE AGAIN?

Traffic light controller examples have often been used in VHDL books and courses for illustrating the behavioral description of an entire system. Indeed, the level of complexity of such a system and the fact that its specification is easy to understand makes it a good candidate for a short case study.

When working down from the specification to the synthesizable description of a simple version of traffic controller, the following problems have to be solved:

- How to describe a unique entity usable at both the specification and synthesizable level?
- How to refine the behavioral description to go to?
- How to deal with initialization?
- How to compare synthesizable and synthesized descriptions?

## 7.2. SPECIFICATION OF THE PROBLEM

The purpose of the traffic lights is to control a crossroad between a farm road and a highway. As long as there is no car on the farm road, priority is given to the main road. The farm road is equipped with car detectors. When at least one car is present on the farm road, and if the main road lights have been green for more than a given time period (LONG_TIME), the main road lights turn to yellow for another given time period (SHORT_TIME). Then, the main road lights turn to red and the farm road lights turn to green at the same time. The farm road lights remain green as long as cars are present on this road but for a maximal time period LONG_TIME. Then, they switch to yellow for a SHORT_TIME and finally to red at the same time as the main road lights become green. The main road lights remain green at least a LONG_TIME and then the cycle is repeated.

Two other functionalities are also provided by the traffic light controller: an economy mode and a high-priority mode.

When the power supply of the crossroad lights fails, the signal ECONOMY is activated, and all the lights of the crossroad instantly switch to flashing red and remain in the same state as long as ECONOMY is activated (sensitive on level). When this signal becomes passive, the regular cycle is therefore resumed: priority is given to the main road whose lights become green for at least a LONG_TIME.

The high priority mode is devoted to high priority vehicles such as firetrucks or ambulances. From their vehicles on the main road, they may send the signal PRIORITY, which is a pulse (i.e., the system is sensitive on one edge).

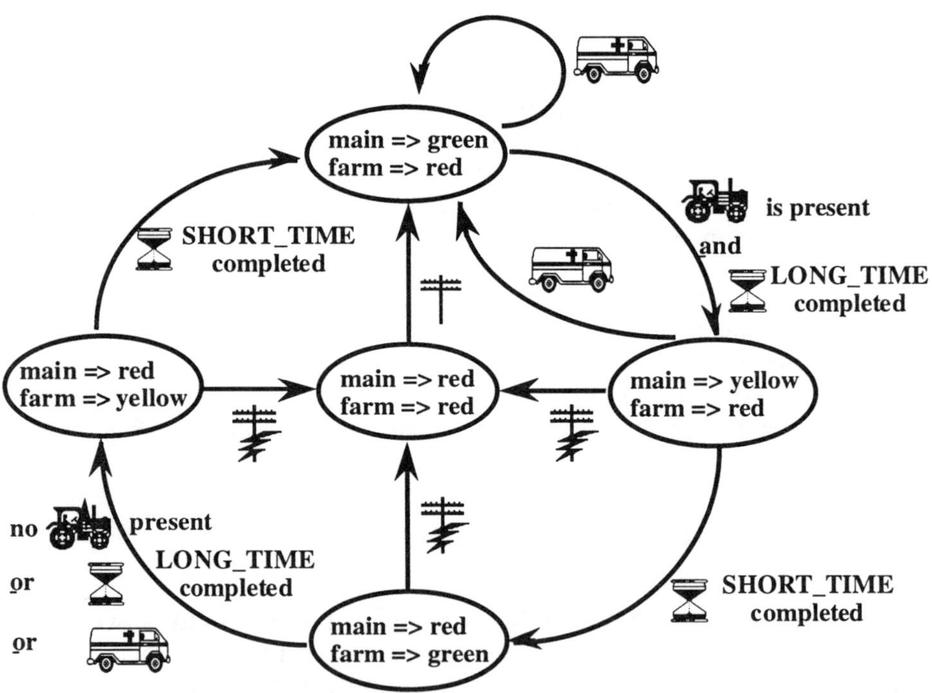


**Figure 7.1. Traffic Light State Diagram**

When received, the following actions are executed before coming back to the regular cycle:
- if the main road lights are green, they will remain green for at least a LONG_TIME after reception of PRIORITY.
- if the main road lights are yellow, they switch to green for at least a LONG_TIME.
- if the main road lights are red, the farm road lights instantly switch to yellow and, after a SHORT_TIME, the main road lights and farm road lights simultaneously switch to green and red respectively. Then, nothing happens for at least a LONG_TIME.

The economy mode, which consists in saving the energy of the battery of the crossroad lights when the power supply fails, takes precedence over the other modes.

Figure 7.1 illustrates theses functionalities with a finite state diagram.


## 7.3. ENTITY   DECLARATION


Modeling from specification down to synthesizable level has some implications on the writing of the entity declaration. A good methodology consists in trying to declare a single entity usable at both specification level and synthesizable level. This allows a simple comparison using the same set of stimuli. This validates the consistency of the refinement process.

Another useful validation consists in comparing the system behavior before and after synthesis: synthesizable towards synthesized descriptions. As data types of ports and generics may be different (integers or enumerated types for synthesizable code and bits or bit-vectors in the synthesized result for example), it is usually necessary to use a conversion function on the ports to keep the same entity declaration.

VHDL offers many powerful features, especially at data type level. They may be used for these descriptions when synthesizable. When these types are user-defined, they have to be declared within a package referenced from the entity:

```
package RESOURCE_PKG is
        type T_DETECTOR is ( CAR, NO_CAR ) ;
        type T_LIGHT is ( FLASHING_RED, MAIN_ROAD_GREEN, MAIN_ROAD_YELLOW,
                                FARM_ROAD_GREEN, FARM_ROAD_YELLOW);
end RESOURCE_PKG ;
```

This package will also be referenced from the testbench units that have to provide the stimuli and check the results.

```
use WORK.RESOURCE_PKG.all ;
entity TRAFFIC is
      generic ( SHORT_TIME      : TIME ;
                LONG_TIME       : TIME );
      port (    CAR_DETECTOR       : in T_DETECTOR ;    -- Dectector on the farm road
                PRIORITY_MODE      : in BOOLEAN ;       -- Signal for ambulances
                ECONOMY_MODE       : in BOOLEAN ;       -- Signal power supply default
                CROSSROAD_LIGHTS : out T_LIGHT);        -- State of the crossroad lights
end TRAFFIC ;
```

## 7.4. DESCRIBING THE BEHAVIORAL ARCHITECTURE

The system description consists of a control part translating the previous
finite state diagram (cf. figure 7.1). The operative part is reduced to timers.

A first glance at the state diagram shows that the "next" state is not unique.
For example, it is possible to go from the state FARM_GREEN to two different
states, depending on conditions. This implies that a description using a single
process with an implicit state variable (i.e., multiple and sequential wait
statements) is not adequate. Therefore, the structure proposed uses an explicit
state variable. This state variable is assigned within the branches of a case
statement to define the "next" state. The description is

```
architecture SPEC_LEVEL of TRAFFIC is
begin

    process
        variable STATE : T_LIGHT := MAIN_ROAD_GREEN ;
    begin
        case STATE is

        when FLASHING_RED =>
            wait until not ECONOMY_MODE;
            STATE := MAIN_ROAD_GREEN;
        when MAIN_ROAD_GREEN =>
            wait on PRIORITY_MODE, ECONOMY_MODE for LONG_TIME;
            if not ECONOMY_MODE then
                if not PRIORITY_MODE'event then   --LONG_TIME is completed
                    if CAR_DETECTOR = NO_CAR then
                        wait on CAR_DETECTOR, PRIORITY_MODE,
                                                ECONOMY_MODE;
                        if ECONOMY_MODE then
                            STATE := FLASHING_RED;
                        elsif CAR_DETECTOR = CAR then
                            STATE := MAIN_ROAD_YELLOW;
                        end if;
```

```
                else
                     STATE := MAIN_ROAD_YELLOW;
                end if;
            end if;
        else
            STATE := FLASHING_RED;
        end if;
    when MAIN_ROAD_YELLOW =>
        wait on PRIORITY_MODE, ECONOMY_MODE for SHORT_TIME;
        if ECONOMY_MODE then
            STATE := FLASHING_RED;
        elsif PRIORITY_MODE'event and PRIORITY_MODE then
            STATE := MAIN_ROAD_GREEN;
        else -- SHORT_TIME is completed
            STATE := FARM_ROAD_GREEN;
        end if;
    when FARM_ROAD_GREEN =>
        wait on PRIORITY_MODE, ECONOMY_MODE, CAR_DETECTOR
                                            for LONG_TIME;
        if ECONOMY_MODE then
            STATE := FLASHING_RED;
        else
            STATE:= FARM_ROAD_YELLOW ;
        end if;

    when FARM_ROAD_YELLOW =>
        wait on ECONOMY_MODE for SHORT_TIME;
        if ECONOMY_MODE then
            STATE := FLASHING_RED;
        else
            STATE := MAIN_ROAD_GREEN;
        end if;
    end case;

    CROSSROAD_LIGHTS <= STATE;
    end process;

end SPEC_LEVEL;
```

**Remark:**
The signals ECONOMY_MODE and PRIORITY_MODE are not processed in the same way. The first forces a state as long as it is active, the second is active when it changes its value from FALSE to TRUE; this is the edge which is important.

## 7.5. DESCRIBING THE SYNTHESIZABLE ARCHITECTURE

### 7.5.1. From Behavioral To Synthesizable Description

Natural hardware resulting from controller synthesis should be a sequential machine. Because it is impossible or too dangerous to infer asynchronous automaton, a clock signal must be introduced as a new input of the system. Indeed synthesis tools cannot provide clock generator. It is therefore the designer who has to insert this signal into the interface to be sensitive to an external clock. At the behavioral level, the time unit needed is the second. If the outside clock period is shorter, it is easy to synthesize hardware so as to divide the frequency compatible with the delay expressions of the controller.

In the behavioral description, which are the nonsynthesizable statements or expressions? Clearly, the difficulty stems from the **wait** statements and the use of the attribute 'EVENT:

- Synchronization on a signal different from the clock is not supported inside a sequential description. The solution is to divide the description into two parts: one sequential and the other combinational.
- Moreover a delay expression (**for** a_time...) is not admitted. A first solution to this problem is for timers to be explicitly introduced within the architecture as components. The load order and the output check to zero belong to the dialogue between state machine and timers. A second solution is to leave the synthesis tool to infer a decrementer by using variables inside the combinational description of the state machine.
- PRIORITY_MODE'EVENT is not synthesizable. It is used in the behavioral description to be sensitive on one edge. The solution is to define a memory point that is set by the edge and reset by the combinational part of the state machine. This memory point can only be a simple latch or, more complex, a flip-flop register synchronized with the clock.

Because all the declared types are enumerated types, they are synthesizable. If this were not the case, the only solution would be to replace nonsupported types by right types for synthesis even if this does not appear obvious.

A generic entity does not directly infer hardware. It must be configured to assign values to generic parameters and finally to freeze the choices authorized by these parameters. Thus to have a hardware result, the example must firstly be instantiated and secondly configured.

To be sure that the state machine works correctly, an initial value must be forced to the state register. So a reset signal may be introduced to set the initial state. This reset can be synchronous or asynchronous (the hardware is smaller in that case). In the controller example, the signal ECONOMY_MODE can provide the reset feature. It is an asynchronous signal, which forces the automaton to FLASHING_RED state. This slight optimization makes one difference from the initial state of the behavioral description, which is MAIN_ROAD_GREEN. To be more consistent between the two descriptions, it would be better to change the initial state of the behavioral description to FLASHING_RED.

## 7.5.2. First Proposition of Synthesizable Architecture

The proposed architecture is composed of several processes: three sequential processes, one concurrent statement, one combinational process and one nonsynthesizable concurrent statement.

The state machine is made of three processes: STATE_REGISTER, NEW_STATE, and TIMERS. The first is devoted to memorizing the current state of the automaton. Therefore the signal CURRENT_STATE that maintains the current value of the state machine, is always assigned to the system outputs CROSSROAD_LIGHTS. The second process is purely combinational; it computes the next state, loads and decrements the timers, and resets the memory point, which has been set by a rising edge of the PRIORITY_MODE signal. The third process loads and decrements the two timers used as watchdogs.

To translate an event of the PRIORITY_MODE signal to a state, a memory point is introduced by the process PRIORITY_MEMO. If the signal ECONOMY_MODE is active, this process resets the local signal PRIORITY in an asynchronous way. In other cases, two input signals PRIORITY_MODE and NO_PRIORITY synchronously set and reset the signal PRIORITY respectively.

Finally to maintain the same behavior as the first architecture, a clock generator is described as a nonsynthesizable concurrent statement. The way to indicate such a property is synthesis tool dependent: it can be an attribute specified on the label of the statement or a portion of code separated from the other synthesizable part by syntactic comments. Of course, at the end, the clock signal will be a new input of the hardware, but to authorize comparisons between behavioral and synthesizable architectures, the clock signal introduction as main input, is delayed to another entity description.

The following schema shows the chosen synthesizable architecture, which is composed of five main processes.

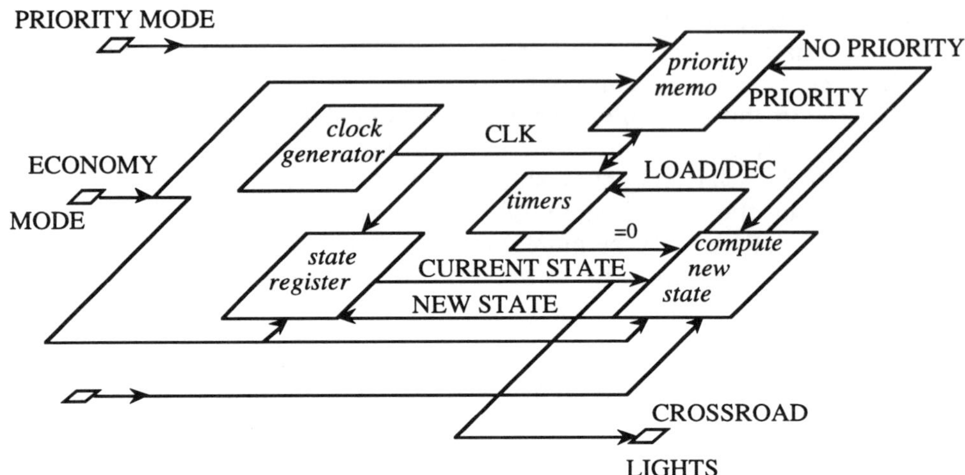


**Figure 7.2. Schematic View of Synthesizable Architecture**

Here is the code of the synthesizable architecture describing the traffic light controller:

```vhdl
architecture SYNTH_LEVEL of TRAFFIC is

    constant PERIOD : TIME := 1 sec;                    -- Clock period
    constant HALF_PERIOD := PERIOD / 2;                 -- Half clock period
    constant LONG_TIME_SEC := LONG_TIME / PERIOD; -- Long_time in clock period units
    constant SHORT_TIME_SEC := SHORT_TIME / PERIOD;

    signal CLK : BIT;                                   -- Local clock signal
    signal CURRENT_STATE, NEW_STATE : T_LIGHT;  -- Intermediate signals coding
                                                        -- current and new state
    signal PRIORITY : BOOLEAN;          -- Memorizes the edge of the external signal
                                        -- PRIORITY_MODE
    signal NO_PRIORITY : BOOLEAN;    -- Priority has been taken into account
    signal LG_TIME : NATURAL range 0 to LONG_TIME;   -- Long timer
    signal SH_TIME : NATURAL range 0 to SHORT_TIME; -- Short timer
    signal LD_LG_TIME, LD_SH_TIME : BOOLEAN;         -- Timer load orders
    signal DEC_LG_TIME, DEC_SH_TIME : BOOLEAN;       -- Timer decrement orders

begin
    -- not synthesizable statement
    CLOCK_GENERATOR : CLK <= not CLK after HALF_PERIOD;

    -- state register to be synthesized
    STATE_REGISTER : process (CLK, ECONOMY_MODE)
    begin
        if ECONOMY_MODE then
```

```vhdl
                CURRENT_STATE <= FLASHING_RED;
        elsif CLK='1' and CLK'event then
                CURRENT_STATE <= NEW_STATE;
        end if;
    end process STATE_REGISTER;


    -- The CROSSROAD_LIGHTS signal reflects the internal values
    -- of CURRENT_STATE signal
    CROSSROAD_LIGHTS <= CURRENT_STATE;


    -- To translate the rising edge of PRIORITY signal into a level
    PRIORITY_MEMO : process (CLK, ECONOMY_MODE)
    begin
        if ECONOMY_MODE then
                PRIORITY <= FALSE;
        elsif CLK='1' and CLK'event then
                if PRIORITY_MODE then
                        PRIORITY <= TRUE;
                elsif NO_PRIORITY then
                        PRIORITY <= FALSE;
                end if;
        end if;
    end process;


    -- Synchronous decrementers as timers
    TIMERS : process (CLK)
    begin
        if LD_LG_TIME then
                LG_TIME <= LONG_TIME_SEC;
        elsif DEC_LG_TIME and (LG_TIME /= 0) then
                LG_TIME <= LG_TIME - 1;
        end if;
        if LD_SH_TIME then
                SH_TIME <= SHORT_TIME_SEC;
        elsif DEC_SH_TIME and (SH_TIME /= 0) then
                SH_TIME <= SH_TIME - 1;
        end if;
    end process ;


    -- Combinational part computing the new state and resetting the memorized prority state
    COMPUTE_NEW_STATE : process (CURRENT_STATE, ECONOMY_MODE,
                                 PRIORITY, CAR_DETECTOR)

    begin
    -- Initial values
        if ECONOMY_MODE then
                NO_PRIORITY <= FALSE;
                NEW_STATE <= FLASHING_RED;
        else
```

```vhdl
-- Default values of signals
        NEW_STATE  <= CURRENT_STATE;
        NO_PRIORITY <= PRIORITY;
        LD_LG_TIME <= FALSE;
        LD_SH_TIME <= FALSE;
        DEC_LG_TIME <= FALSE;
        DEC_SH_TIME <= FALSE;

        case CURRENT_STATE is

        when FLASHING_RED =>
            NEW_STATE <= MAIN_ROAD_GREEN;
            LD_LG_TIME <= TRUE;

        when MAIN_ROAD_GREEN =>
            if PRIORITY then
                LD_LG_TIME <= TRUE;
            elsif LG_TIME = 0 then
                if CAR_DETECTOR = CAR then
                    NEW_STATE <= MAIN_ROAD_YELLOW;
                    LD_SH_TIME <= TRUE;
                end if;
            else
                DEC_LG_TIME <= TRUE;
            end if;

        when MAIN_ROAD_YELLOW =>
            if PRIORITY then
                NEW_STATE <= MAIN_ROAD_GREEN;
                LD_LG_TIME <= TRUE;
            elsif SH_TIME = 0 then
                NEW_STATE <= FARM_ROAD_GREEN;
                LD_LG_TIME <= TRUE;
            else
                DEC_SH_TIME <= TRUE;
            end if;

        when FARM_ROAD_GREEN =>
            if LG_TIME = 0 then
                NEW_STATE <= FARM_ROAD_YELLOW ;
                LD_SH_TIME <= TRUE;
            else
                DEC_LG_TIME <= TRUE;
            end if;

        when FARM_ROAD_YELLOW =>
            if SH_TIME = 0 then
                NEW_STATE <= MAIN_ROAD_GREEN;
                LD_LG_TIME <= TRUE;
            else
                DEC_SH_TIME <= TRUE;
```

```
                    end if;
                  end case ;
              end if;
          end process;

end SYNTH_LEVEL ;
```

## 7.5.3 Second Proposition of Synthetizable Architecture

This second architecture replaces the synthetizable processes of the preceding proposition by only one that remains synthetizable. The resulting hardware deduced by the synthesis tool and not explicitly described by the designer is the same as for the first architecture. Here is the code of the new architecture containing a main process, which entirely described the final state machine behavior with the exception of clock generator.

```
architecture SECOND_SYNTH_LEVEL of TRAFFIC is

    constant PERIOD : TIME := 1 sec;                       -- Clock period
    constant HALF_PERIOD := PERIOD / 2;                    -- Half clock period
    constant LONG_TIME_SEC := LONG_TIME / PERIOD; -- Long_time in clock period units
    constant SHORT_TIME_SEC := SHORT_TIME / PERIOD;

    signal CLK : BIT;                        -- Local clock signal
    signal PRIORITY : BOOLEAN;               -- Memorizes the edge of the external signal
                                             -- PRIORITY_MODE

begin
    -- Not synthesizable statement
    CLOCK_GENERATOR : CLK <= not CLK after HALF_PERIOD;

    -- Process describing the Final State Machine and the operative part
    FSM : process (CLK, ECONOMY_MODE)
        variable STATE : T_LIGHT;                                -- State register
        variable LG_TIME : NATURAL range 0 to LONG_TIME;   -- Long timer
        variable SH_TIME : NATURAL range 0 to SHORT_TIME;  -- Short timer

    begin
        if ECONOMY_MODE then                  -- Asynchronous part
            STATE := FLASHING_RED;

        elsif CLK'event and CLK='1' then      -- Synchronous part
            if PRIORITY_MODE then
                PRIORITY <= TRUE;
            end if;
            case STATE is
            when FLASHING_RED =>
```

```vhdl
                    STATE := MAIN_ROAD_GREEN;
                    LG_TIME := LONG_TIME_SEC;

                when MAIN_ROAD_GREEN =>
                    if PRIORITY then
                        LG_TIME := LONG_TIME_SEC;
                        PRIORITY <= FALSE;
                    elsif LG_TIME = 0 then
                        if CAR_DETECTOR = CAR then
                            STATE := MAIN_ROAD_YELLOW;
                            SH_TIME := SHORT_TIME_SEC;
                        end if;
                    else
                        LG_TIME := LG_TIME -1;
                    end if;

                when MAIN_ROAD_YELLOW =>
                    if PRIORITY then
                        STATE := MAIN_ROAD_GREEN;
                        LG_TIME := LONG_TIME;
                        PRIORITY <= FALSE;
                    elsif SH_TIME = 0 then
                        STATE := FARM_ROAD_GREEN;
                        LG_TIME :=LONG_TIME_SEC;
                    else
                        SH_TIME := SH_TIME - 1;
                    end if;

                when FARM_ROAD_GREEN =>
                    if LG_TIME = 0 then
                        STATE := FARM_ROAD_YELLOW ;
                        SH_TIME := SHORT_TIME_SEC;
                    else
                        LG_TIME <= LG_TIME - 1;
                    end if;

                when FARM_ROAD_YELLOW =>
                    if SH_TIME = 0 then
                        STATE := MAIN_ROAD_GREEN;
                        LG_TIME := LONG_TIME_SEC;
                    else
                        SH_TIME := SH_TIME - 1;
                    end if;
                end case ;
            end if;
        -- The signal CROSSROAD_LIGHTS reflects the internal values
        -- of the variable STATE
        CROSSROAD_LIGHTS <= STATE;
        end process;
end SECOND_SYNTH_LEVEL;
```

This architecture is more straightforward than the first synthesizable one. Many intermediate signals (such as DEC_SH_TIME, NO_PRIORITY, NEW_STATE, ...) have disappeared and are replaced by local variables. Indeed, this description is very close to the first behavioral description given in this chapter (section 7.4.). The only difference consists in introducing a clock signal and constitutes the real difficulty of the transformation to obtain a synthesizable code.

**Remark:**
The signal PRIORITY is set and tested in the same process (and also in the same clock period), however the set action is only taken in account after one period delay. It would be very dangerous to put a variable and not a signal to memorize the PRIORITY activity. If a variable was used instead of a signal, the hardware resulting from the synthesis would not be consistent with simulation before synthesis. Anyway, if a signal or a variable is used to memorize PRIORITY, the synthesis provides a register and furthermore, a delay of one period is introduced between set and test.

## 7.6. DESIGNER'S CONCERNS

### 7.6.1. Tool Dependency

Concerning tool dependency, one good point is that VHDL code can be the same for many synthesis tools and this example is well accepted by several of them. However the way to define a nonsynthesizable part of code is tool dependent. Moreover the constraint descriptions differ from one tool to another. Their syntax, and also their semantics are defined for only one synthesis context. In fact, except for the architecture description written in VHDL, all information concerning the hardware context is described according to the interface provided by the synthesis tool. For example, if ports defined by the entity at the top level of hierarchy have to become real inputs/outputs of the circuit, special hardware must be provided. This kind of information is described to the synthesis tool in its own language.

### 7.6.2. Constraint Checking

When the desired hardware is synchronous, the main constraint to be respected by the synthesis process is the period of the clock. After describing the electrical characteristics of the ports (strength to drive, capacitance), the synthesis tool may accept the period to be respected as a constraint. It is not always the case; some tools do not support this data, and the designer has to check the critical path delays. If the constraints cannot be respected, the only solution, for a given library, is to choose another architecture to find improved optimizations. Consequently a new VHDL code must be written and then synthesized. If all constraints are accepted, the hardware result of synthesis must be the smallest in terms of area.

# 8. APPENDIX

## 8.1. GRAMMAR  SUMMARY

This grammar is not a full VHDL grammar; it is described here to save the designer from referring the LRM.

The new syntax rules introduced by *VHDL'92* are underlined. [x] means that x is optional but when it appears, it appears only once. {x} means that x is optional but may appear several times. x | y means that either x or y can appear.

### Design unit declarations

*   **entity** identifier **is**
                [ **generic** ( generic_list ) ] [ **port** ( port_list ) ; ]
                { entity_declarative_item }
    [begin
                {concurrent_assertion_statement
                | *passive*_concurrent_procedure_call
                | *passive*_process_statement}]
    **end** [ <u>entity</u> ] [ *entity*_simple_name ] ;

*   **architecture** identifier **of** *entity*_name **is**
                { architecture_declarative_item }
    begin
                { concurrent_statement }
    **end** [ <u>architecture</u> ] [ *architecture*_simple_name ] ;

*   **package** *package*_simple_name **is**
                { package_declarative_item }
    **end** [ <u>package</u> ] [ *package*_simple_name ] ;

*   [ **package body** *package*_simple_name **is**
                { package_body_declarative_item }
    **end** [ <u>package body</u> ] [ *package*_simple_name ] ; ]

*   **configuration** identifier **of** *entity*_name **is**
                { use_clause | attribute_specification | <u>group specification</u> }
                **for** *architecture*_name | *block_statement*_label
                    | *generate_statement*_label [ ( index_specification ) ]
                    { use_clause }
                    { block_configuration | component_configuration }
                **end for** ;
    **end** [ <u>configuration</u> ] [ *configuration*_simple_name ] ;

## Sequential Statements

- [ label : ] **assert** condition [ **report** expression ] [ **severity** expression ]
- [ label : ] **report** expression [ **severity** expression ] ;
- [ label : ] **wait** [ **on** signal_name { ,signal_name } ] [ **until** condition ] [ **for** time_expression ] ;
- [ label : ] target := expression ;
- [ label : ] target <= [ **transport** | [ **reject** time_expression ] **inertial** ] waveform ;
- [ if_label : ] **if** condition **then**
                        sequential_statement { sequential_statement }
                { **elsif** condition **then**
                        sequential_statement { sequential_statement }
                [ **else**
                        sequential_statement { sequential_statement } ]
                **end if** [ if_label ] ;
- [ case_label : ] **case** expression **is**
                        **when** choices => sequential_statement { sequential_statement }
                        { **when** choices => sequential_statement { sequential_statement }}
                        **end case** [ case_label ] ;
- [ loop_label : ] [ **while** condition | **for** loop_parameter_specification ] **loop**
                        sequential_statement;  { sequential_statement; }
                        **end loop** [ loop_label ] ;
- [ label : ] procedure_name [ ( actual_parameter_part ) ] ;
- [ label : ] **next** [ loop_label ] [ **when** condition ] ;
- [ label : ] **exit** [ loop_label ] [ **when** condition ] ;
- [ label : ] **return** [ expression ] ;
- [ label : ] **null** ;

## Subprogram declarations

- **procedure** procedure_name [ ( [10]interface_element { ; interface_element } ) ] ;
- [ **pure** | **impure** ] **function** function_name
                [ ( interface_element { ; interface_element } ) ] **return** type_mark ;
- **procedure** procedure_name [ ( interface_element { ; interface_element } ) ] **is**
                { subprogram_declarative_item }
**begin**
                sequential_statement { sequential_statement }
**end** [ **procedure** ] [ procedure_name ] ;
- [ **pure** | **impure** ] **function** function_name
                [ ( interface_element { ; interface_element } ) ] **return** type_mark **is**
                { subprogram_declarative_item }
**begin**
                sequential_statement { sequential_statement }
**end** [ **function** ] [ function_name ] ;

---

10  interface_element ::= interface_constant_declaration | interface_signal_declaration
                        | interface_variable_declaration | interface_file_declaration
interface_variable_declaration ::=
        [**variable**] identifier_list : [ mode ] subtype_indication [ := static_expression ]
interface_constant_declaration ::=
        [ **constant** ] identifier_list : [ **in** ] subtype_indication [ := static_expression ]
interface_file_declaration ::= **file** identifier_list; [ mode ] subtype_indication
interface_signal_declaration ::=
    [**signal**] identifier_list : [ mode ] subtype_indication [ **bus** ] [ :=static_expression ] ]

## Concurrent Statements

- [ process_label : ] [ **postponed** ] **process** [ ( signal_name { , signal_name } ) ] [ **is** ]
                          process_declarative_part
                **begin**
                      { sequential_statement }
                **end** [ **postponed** ] **process** [ process_label ] ;
- block_label : **block** [ ( guard_expression ) ] [ **is** ]
                [ **generic** ( generic_list ) [ **generic map** ( generic_association_list ) ; ] ]
                [ **port** ( port_list ) [ **port map** ( port_association_list ) ; ] ]
                  [ { block_declarative_item } ]
                **begin** ]
                      { concurrent_statement }
                **end block** [ block_label ] ;
- **with** expression **select**
  target  <= [ **guarded** ] [ **transport** I [ **reject** time_expression ] **inertial** ]
                          { waveform **when** choices , } waveform **when** choices ;
- instantiation_label :  [11]instantiated_unit
                      [ **generic map** ( generic_association_list ) ]
                      [ **port map** ( port_association_list ) ] ;
- [ label : ] [ **postponed** ] **assert** condition [**report** expression ] [**severity** expression];
- [ label : ] [ **postponed** ]  procedure_name [ ( actual_parameter_part ) ] ;
- [ label : ] [ **postponed** ] target  <=
                      [ **guarded** ] [ **transport** I [ **reject** time_expression ] **inertial** ]
                      { waveform **when** condition **else** } waveform [ **when** condition ] ;
- [ label : ] [ **postponed** ] selected_signal_assignment
- generate_label : **for** generate_parameter_specification I **if** condition
                **generate**
                      [ { block_declarative_item } ]
                **begin** ]
                      { concurrent_statement }
                **end generate** [ generate_label ] ;

## Clauses

- **library** logical_name { , logical_name } ;
- **use** logical_name .[12]suffix { ,logical_name.[3] suffix } ;

## Object Declarations

- **constant** identifier { , identifier } : subtype_indication [ := expression ] ;
- [ **shared** ] **variable** identifier { , identifier } : subtype_indication [ := expression ] ;
- **signal** identifier{ , identifier} : subtype_indication [ **register** I **bus**] [ := expression ] ;
- **file** identifier_list: [ mode ] subtype_indication;

---

11  Instantiated_unit ::= [ **component** ] component_name
    I **entity** entity_name [ ( architecture_identifier ) ] I **configuration** configuration_name

12 suffix ::= simple_name I simple_name.all I simple_name.simple_name I simple_name.operator_symbol I  **all**

## Subtype and Type Declarations

* **subtype** identifier **is** [ resolution_function_name ] type_mark [ constraint ] ;
* **type** identifier ;
* **type** identifier **is** ( enumeration_literal { , enumeration_literal } ) ;
* **type** identifier **is** range_constraint **units**
                base_unit_declaration { secondary_unit_declaration }
            **end units** [ physical type simple name ] ;
* **type** identifier **is array** index_constraint **of** element_subtype_indication ;
* **type** identifier **is array** ( type_mark **range** <> { , type_mark **range** <> } )
                                    **of** element_subtype_indication ;
* **type** identifier **is record**
                element_declaration { element_declaration }
    **end record** [ record type simple name ] ;
* **type** identifier **is file** **of** type_mark ;

## Miscellaneous Declarations

* **attribute** identifier : type_mark ;
* **attribute** attribute_designator **of** entity_specification **is** expression ;
* **group** identifier **is** ( entity class entry list ) ;
* **group** identifier : group type name ( group constituent { , group constituent } ) ;
* **alias** [13]alias_designator [ : subtype_indication ] **is** name [ signature[14] ] ;
* **component** identifier [ **is** ]
                [ **generic** ( generic_list ) ] [ **port** ( port_list ) ; ]
    **end component** [ component simple name ] ;
* **file** identifier { , identifier } : subtype_indication **is**
                [ **open** file_open_kind_expression ] **is** file_logical_name ;

## Miscellaneous Specifications

* **for** instantiation_list : component_name
                **use** [ **entity** entity_name [ ( architecture_identifier) ]
                                    | **configuration** configuration_name | **open** ]
                [ **generic map** ( generic_association_list ) ]
                [ **port map** (port_association_list) ] ;
* **disconnect** guarded_signal_specification **after** time_expression ;

---

13 alias_designator := identifier | character literal | operator symbol

14 signature ::= ( [ type_mark { , type_mark } ] [ **return** type_mark ] )

## 8.2. MEMO

### 8.2.1. Process: Is Inferred Hardware Combinational or Sequential?

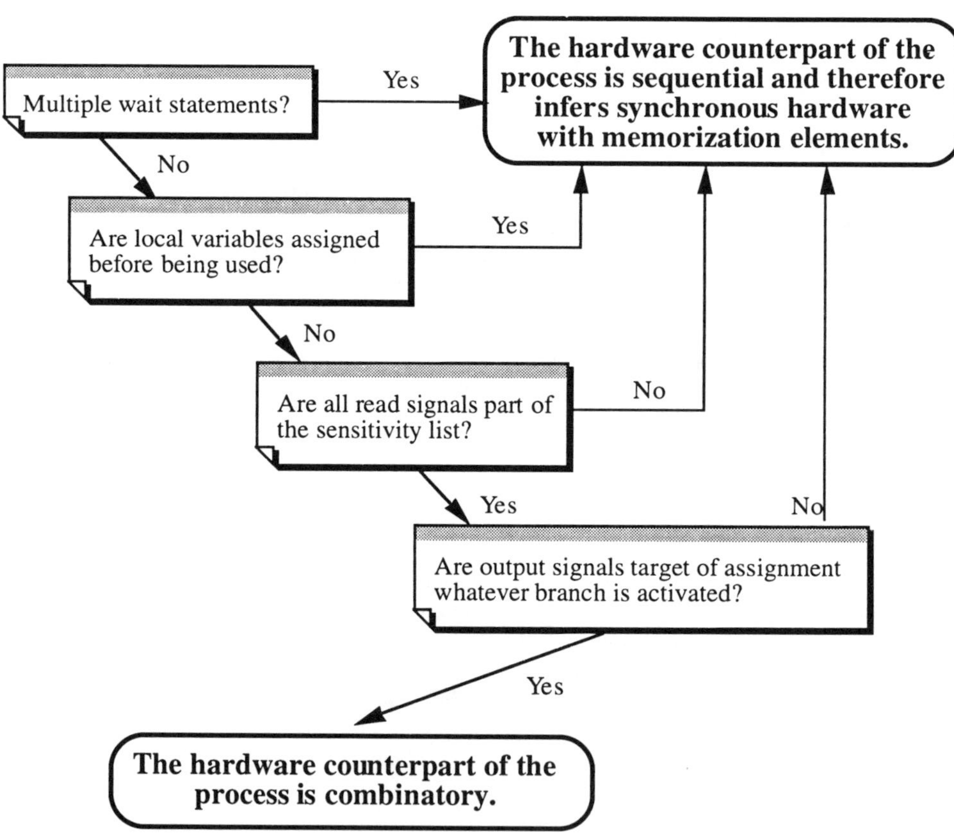

## 8.2.2. Combinational Circuits: Logic Gates

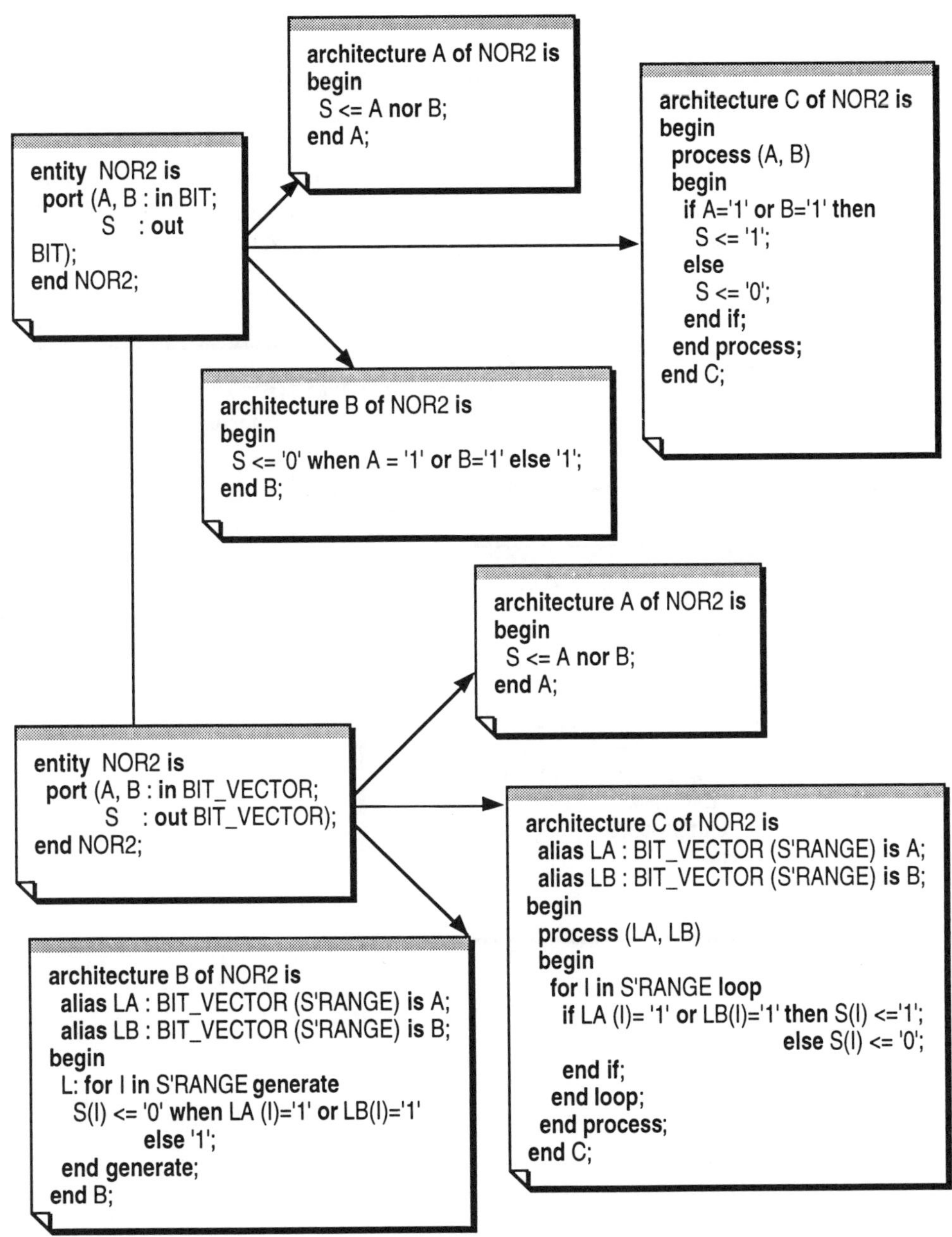

## 8.2.3. Combinational Circuits: Multiplexers

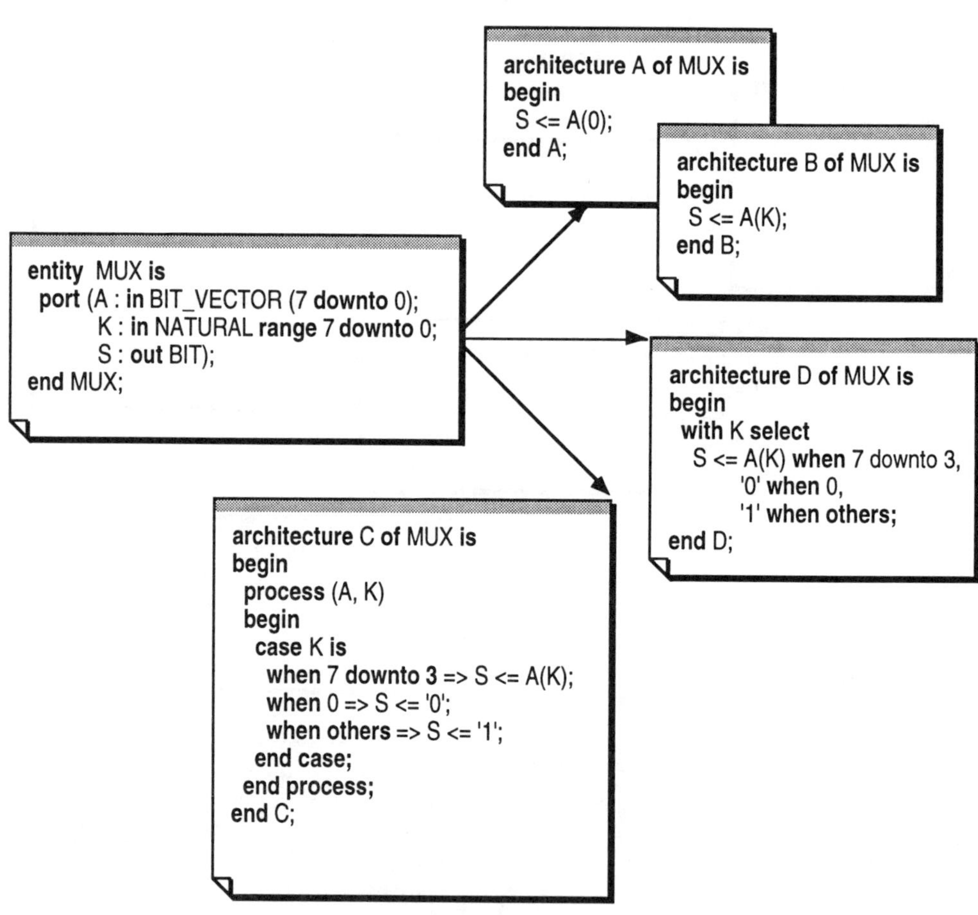

## 8.2.4. Combinational Circuits: Three-State Operators

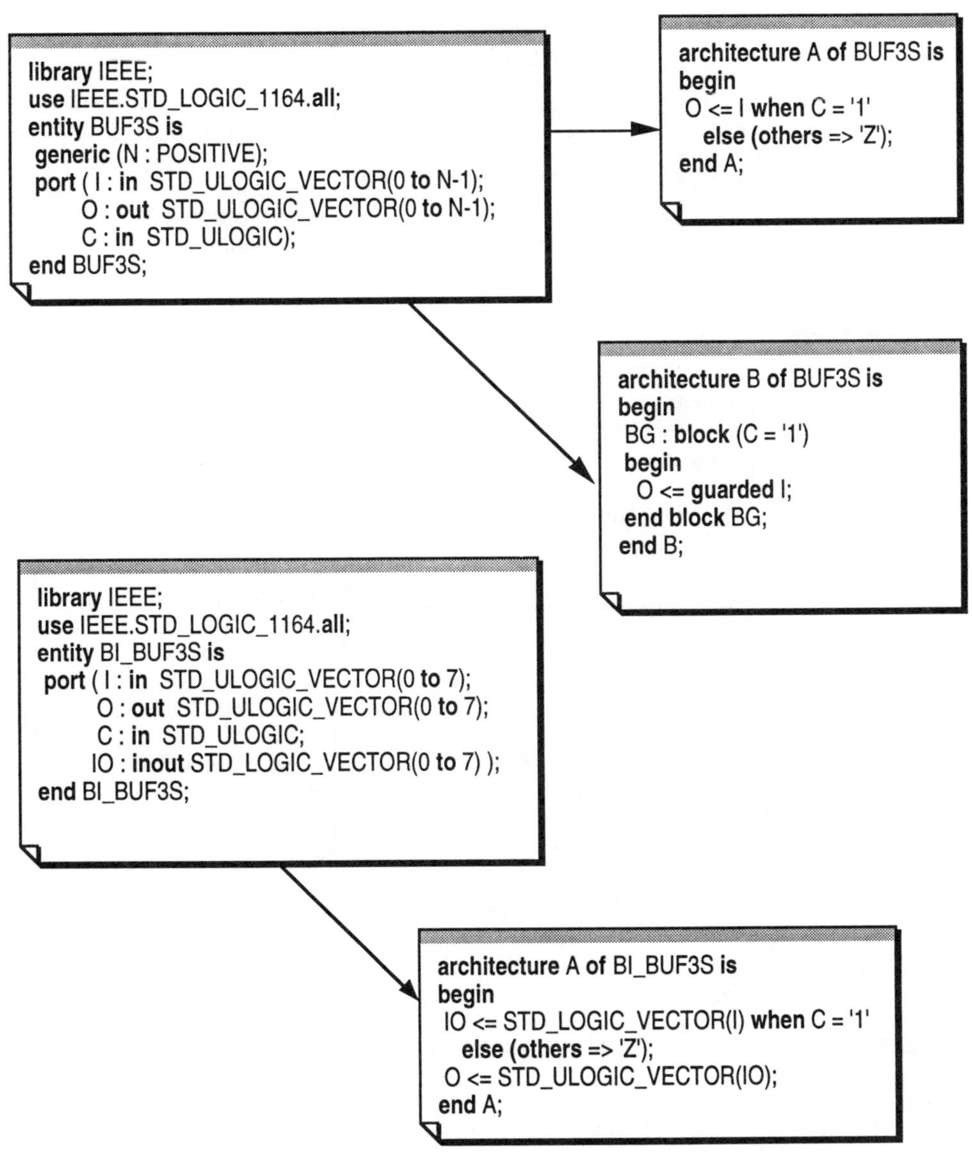

## 8.2.5. Sequential Circuits: Latch and Register

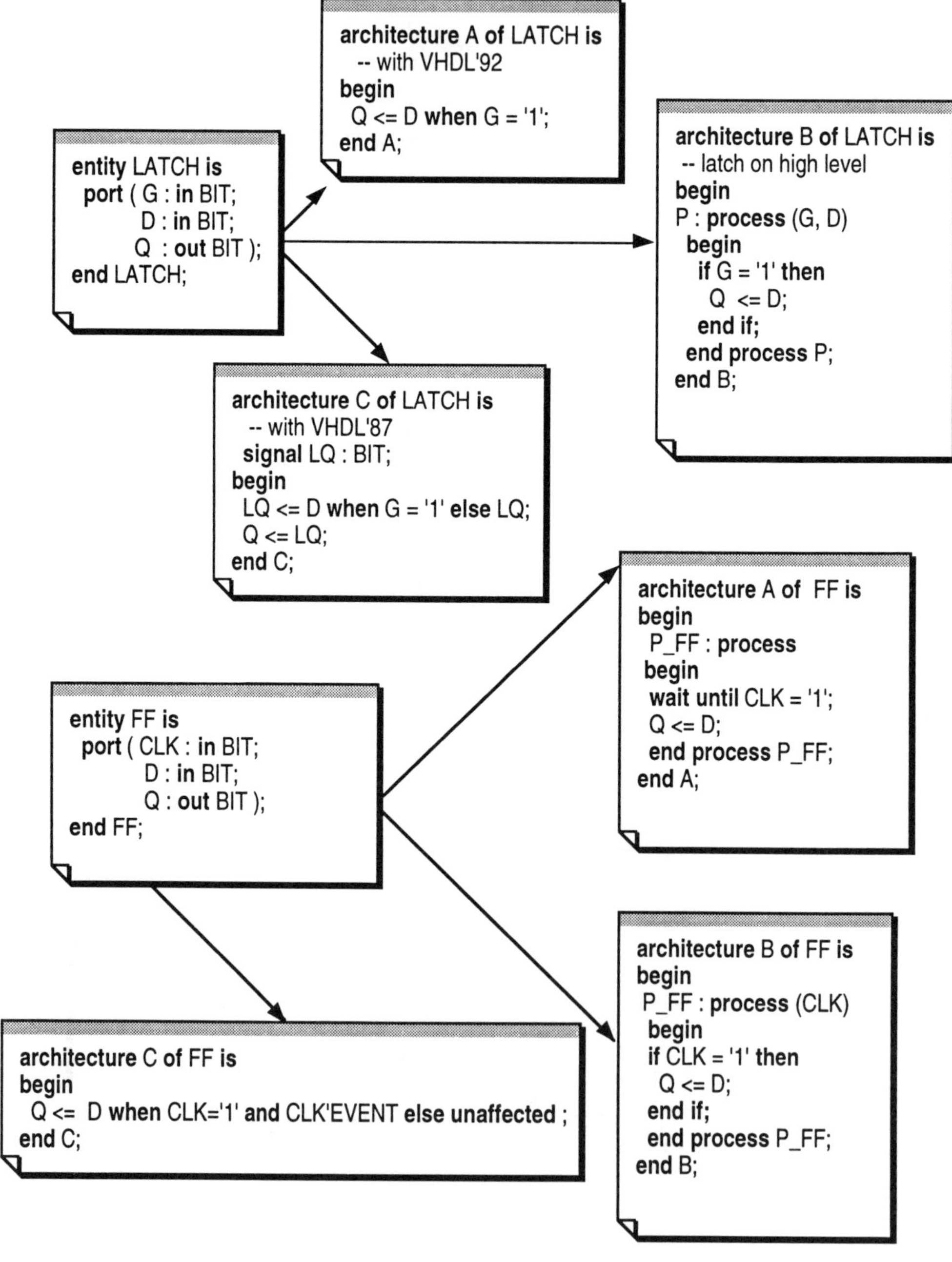

## 8.2.6. Sequential Circuits: Counter with Reset

```
entity COUNTER is
  generic (N : NATURAL);
  port ( CK : in BIT;
         RESET : in BIT;
         R : out NATURAL range 0 to N-1);
end COUNTER;
```

```
architecture SYNC of COUNTER is
  signal C : NATURAL range 0 to N -1;
begin
 R <= C ;
 P_COUNT : process (CK )
 begin
  if CK='1' and CK'EVENT then
   if RESET=1 or C=N-1 then
    C <= 0;
   else
    C <= C + 1;
   end if;
  end if;
 end process P_COUNT;
end SYNC;
```

```
architecture ASYNC of COUNTER is
  signal C : NATURAL range 0 to N -1;
begin
 R <= C ;
 P_COUNT : process (CK )
 begin
 if RESET ='1' then
  C <= 0;
 elsif CK='1' and CK'EVENT then
  if C=N-1 then
   C <= 0;
  else
   C <= C + 1;
  end if;
 end if;
 end process P_COUNT;
end ASYNC;
```

## 8.2.7. Sequential  Circuits:  Finite  State  Machine

```
entity FSM is
  port ( COND1, COND2 : in BOOLEAN ;
         CK : in BIT ; INIT : in BIT;
         OUTPUT1, OUTPUT2, OUTPUT3 : out BIT );
end FSM;
```

```
architecture B of  FSM is
  subtype ST_TYPE is NATURAL range 0 to 2
  signal STATE, NEW_STATE : ST_TYPE;
begin
 P_STATE : process  (INIT, CK)
 begin
 if  INIT = '1' then
   STATE <= 0;
 elsif CK='1' and CK'EVENT then
   STATE <= NEW_STATE;
 end if ;
 end process P_STATE;

P_FSM:process (STATE,COND1,COND2)
begin
 OUTPUT1<= '0'; OUTPUT2 <= '0';
 OUTPUT3 <= '0';
 case STATE is
   when 0 => OUTPUT1 <= '1';
             if COND1 then NEW_STATE<=
   when 1 => OUTPUT2 <= '1';
             NEW_STATE <= 0;
   when 2 => OUTPUT3 <= '1';
             if COND2 then NEW_STATE<= 1; end if ;
 end case ;
end process P_FSM;
end B;
```

```
architecture A of FSM is
begin
 P : process
   type STATE_TYPE is (S1, S2, S3) ;
   variable STATE : STATE_TYPE;
begin
 if INIT = '1' then
   STATE := S1;
 elsif CK'EVENT and CK = '1' then
   OUTPUT1 <= '0'; OUTPUT2 <= '0';
   OUTPUT3 <= '0';
   case STATE is
     when S1 => OUTPUT1 <= '1';
                if COND1 then
                  STATE := S3;
                end if;
     when S2 => OUTPUT2 <= '1';
                STATE := S1;
     when S3 => OUTPUT3 <= '1';
                if COND2 then
                  STATE := S2;
                end if;
   end case;
 end if;
 wait on CK, INIT;
 end process P;
end A;
```

## 8.3 INDEX